Resource Allocation with Carrier Aggregation in Cellular Networks

Haya Shajaiah • Ahmed Abdelhadi
Charles Clancy

Resource Allocation with Carrier Aggregation in Cellular Networks

Optimality and Spectrum Sharing using C++ and MATLAB

Springer

Haya Shajaiah
Virginia Tech
Arlington, VA, USA

Ahmed Abdelhadi
Virginia Tech
Blacksburg, VA, USA

Charles Clancy
Virginia Tech
Arlington, VA, USA

ISBN 978-3-319-60539-5 ISBN 978-3-319-60540-1 (eBook)
DOI 10.1007/978-3-319-60540-1

Library of Congress Control Number: 2017943167

Printed on acid-free paper

This Springer imprint is published by Springer Nature
The registered company is Springer International Publishing AG
The registered company address is: Gewerbestrasse 11, 6330 Cham, Switzerland

To my parents; my beloved husband, Islam; and our beloved children Rasem, Tayma, and Yara.

Preface

Recently, there has been a massive growth in the number of mobile users and their traffic, with the data traffic volume almost doubling every year but users are very limited to the service providers' resources. Increasing the utilization of the existing spectrum can significantly improve network capacity, data rates, and user experience. Spectrum sharing enables wireless systems to harvest under utilized swathes of spectrum, which vastly increases the efficiency of spectrum usage. Making spectrum more widely available can provide significant gain in mobile broadband capacity only if those resources can be aggregated efficiently with the existing commercial mobile system resources. Carrier aggregation (CA) is one of the most distinct features of 4G systems including Long Term Evolution-Advanced (LTE-Advanced). This volume introduces an efficient resource management approach for future spectrum sharing systems. The book focusses on providing optimal resource allocation framework based on carrier aggregation to allocate multiple carriers' resources efficiently among mobile users with elastic and inelastic traffic in cellular networks. The allocation policy is based on utility proportional fairness, where the fairness among users is in utility percentage of the application running on the user equipment (UE). Resource allocation (RA) with CA is proposed to allocate single or multiple carriers' resources optimally among users subscribing for mobile services. Each user is guaranteed a minimum quality of service (QoS) that varies based on the user's application type. Furthermore, it provides an optimal traffic-dependent pricing mechanism that could be used by network providers to charge mobile users for the allocated resources. The book provides different resource allocation with carrier aggregation solutions, for different spectrum sharing scenarios, and compares them. The provided solutions consider the diverse quality of experience requirement of multiple applications running on the users' equipment since different applications require different application performance. In addition, the book addresses the resource allocation problem for spectrum sharing systems that require user discrimination when allocating the network resources. Furthermore, an application-aware resource block (RB) scheduling with CA is proposed to assign RBs of multiple component carriers to users' applications based on a utility proportional fairness scheduling policy.

Arlington, VA, USA Haya Shajaiah
Blacksburg, VA, USA Ahmed Abdelhadi
Arlington, VA, USA Charles Clancy
April 2017

Acknowledgments

I would like to thank my advisor Dr. T. Charles Clancy for all his support, advice, help, and guidance. I would not have accomplished this work without his vision and support. I also thank Dr. Ahmed Abdelhadi, currently research professor at Virginia Tech, for being a very helpful mentor during my PhD studies and for co-authoring this book with me. I thank all my family members for their support and encouragement. I would like to express my gratitude to my beloved husband, Dr. Islam Younis, for his constant love, support, and encouragement. I would have to admit that this work would not have been possible without Islam's support and love. I would also like to express my gratitude to my parents, Dr. Jamal Nouh Shajaiah and Amal Okab, for their encouragement, support, unconditional love, and care. My dad has always been my role model in life. I admire and regard him with utmost respect. My mom has done a lot for me; I specially thank her for all that she has done for me. I love my parents so much, and I would not have made it this far without them. I also thank my sisters, Dr. Hiba and Hala, and my brother, Hussein, for their continuous encouragement and support.

Arlington, VA, USA
April 2017

Haya Shajaiah

Contents

1 Introduction 1
1.1 Motivation and Background 1
1.2 Carrier Aggregation 3
1.2.1 Motivation for Developing Carrier Aggregation with a QoE approach 3
1.2.2 Deployment Scenarios for Carrier Aggregation 4
1.2.3 Types of Carrier Aggregation 6
1.3 Previous Studies in Resource Allocation for Spectrum Sharing 7
1.4 Overview 10
1.5 Book Organization 11
References 12

2 Utility Functions and Resource Allocation for Spectrum Sharing 19
2.1 User Applications Utility Functions 19
2.2 Utility Proportional Fairness Resource Allocation 20
2.3 Utility Proportional Fairness Resource Allocation with Carrier Aggregation 21
2.4 MATLAB Code 22
References 23

3 Multi-Stage Resource Allocation with Carrier Aggregation 25
3.1 Distributed Resource Allocation with Carrier Aggregation 25
3.1.1 Problem Formulation 26
3.1.2 Primary and Secondary Carriers Optimization Problem 27
3.1.3 Multi-Stage Distributed Resource Allocation Algorithm 30
3.1.4 Optimal Rate Allocation Numerical Results 31
3.2 Centralized Resource Allocation with Carrier Aggregation 38
3.2.1 Multi-Carrier System Model 38
3.2.2 Multi-Carrier Optimization Problem 39
3.2.3 A Price Selective Centralized RA with CA Algorithm 42
3.2.4 Multi-Carrier Rate Allocation Numerical Results 45
3.3 Summary and Conclusion 48

3.4 C++ Codes ... 48
3.4.1 First Code ... 49
3.4.2 Second Code ... 55
References ... 61

4 Resource Allocation with User Discrimination for Spectrum Sharing ... 63
4.1 Spectrum Sharing Between Public Safety and Commercial Users in Cellular Networks ... 64
4.1.1 Spectrum Sharing Problem Formulation ... 65
4.1.2 Spectrum Sharing Optimization ... 66
4.1.3 Algorithm for Optimal RA with Spectrum Sharing ... 68
4.1.4 Spectrum Sharing Simulation Results ... 69
4.2 Multi-Application Resource Allocation with User Discrimination in Cellular Networks ... 75
4.2.1 Problem Formulation ... 76
4.2.2 Resource Allocation Optimization Problem ... 77
4.2.3 Resource Allocation with User Discrimination Algorithms ... 80
4.2.4 Numerical Results ... 83
4.2.5 Price Sensitivity to Change in *R* ... 86
4.2.6 Price Sensitivity to Change in the Number of UEs ... 87
4.3 Resource Allocation with User Discrimination for Multi-Carrier Cellular Networks ... 88
4.3.1 Multi-Carrier Problem ... 89
4.3.2 Multi-Carrier RA with User discrimination Optimization Problem ... 91
4.3.3 Multi-Carrier Optimization Algorithm ... 98
4.3.4 Numerical Results ... 100
4.4 Summary and Conclusion ... 105
4.5 C++ Codes ... 106
4.5.1 First Code ... 106
4.5.2 Second Code ... 112
4.5.3 Third Code ... 116
4.5.4 Fourth Code ... 122
References ... 130

5 Resource Allocation with Carrier Aggregation for Commercial Use of 3.5 GHz Spectrum ... 133
5.1 Spectrum Sharing Problem Formulation ... 134
5.2 Resource Allocation Optimization for Spectrum Sharing with the 3.5 GHz Spectrum ... 135
5.3 The Macro Cell and Small Cells RA Optimization Algorithm ... 138
5.4 Spectrum Sharing Simulation Results ... 139
5.4.1 Small Cell Allocated Rates and Users QoE ... 140
5.4.2 Macro Cell Allocated Rates and Users QoE ... 142
5.5 Summary and Conclusion ... 142
5.6 C++ Code ... 144
References ... 150

6 RA with CA for a Cellular System Sharing Spectrum with S-Band Radar 153
6.1 System Model 154
6.2 Radar-LTE Spectrum Sharing Approach 156
6.3 Spectrum Sharing Algorithms 156
6.3.1 Channel-Selection Algorithm 157
6.3.2 Null-Space Projection (NSP) Algorithm 157
6.4 RA with CA for Radar-LTE Spectrum Sharing 159
6.5 Two-Stage Carrier Aggregation Algorithm 161
6.6 Numerical Results 163
6.6.1 Gain in LTE-Advanced Cellular Network Resources Due to Spectrum Sharing with MIMO Radar 164
6.6.2 Rate Allocation for $10 \leq R_{\mathrm{LTE}} \leq 70$ in the First Stage of the RA Algorithm 165
6.6.3 Rate Allocation for $10 \leq R_{\mathrm{radar}} \leq 80$ in the Second Stage of the RA Algorithm 167
6.6.4 RA with Carrier Aggregation for $10 \leq R \leq 150$ 168
6.6.5 Price Sensitivity to Change in R 169
6.7 Summary and Conclusion 169
References 170

7 Utility Proportional Fairness Resource Block Scheduling with Carrier Aggregation 173
7.1 System Model and Problem Setup 174
7.2 User Grouping Method 175
7.3 RB Scheduling with CA Problem 176
7.4 Centralized UPFRS-CA Optimization Algorithm 179
7.5 Numerical Results 180
7.6 Summary and Conclusion 183
7.7 MATLAB Code 183
References 186

8 Conclusion and Future Trajectory 189
References 190

Appendix: C++ Functions 193

Index 199

Acronyms

3GPP	3rd generation partnership project
4G	Fourth generation
5G	Fifth generation
BS	Base station
CA	Carrier aggregation
CC	Component carrier
CSI	Channel state information
eNodeB	Evolved node B
FFR	Fractional frequency reuse
FTP	File transfer protocol
Hz	Hertz
IMT-Advanced	International Mobile Telecommunications-Advanced
LTE	Long term evolution
LTE-A	LTE-Advanced
MIMO	Multi input multi output
NBP	National broadband plan
NPSTC	National Public Safety Telecommunications Council
NSP	Null-space projection
NTIA	National Telecommunications and Information Administration
OFV	Objective function value
PF	Proportional fairness
QoE	Quality of experience
QoS	Quality of service
RA	Resource allocation
RB	Resource block
RF	Radio frequency
RRH	Remote radio head
SNR	Signal to noise ratio
SVD	Singular value decomposition

UE	User equipment
UMTS	Universal mobile terrestrial system
UPF	Utility proportional fairness
UPFRS-CA	Utility proportional fairness resource scheduling with carrier aggregation
VoIP	Voice-over-IP

Chapter 1
Introduction

In recent years, the number of mobile subscribers and their traffic has increased rapidly. Mobile subscribers are currently running multiple applications, simultaneously, on their smart phones that require a higher bandwidth and make users so limited to the carrier resources. Multiple services are now offered by network providers such as mobile-TV and multimedia telephony [1]. According to the Cisco Visual Networking Index (VNI) [2], the volume of data traffic is expected to continue growing up and reaches 1000 times its value in 2010 by 2020 which is referred to as 1000× data challenge. With the increasing volume of data traffic, more spectrum is required [3]. However, due to spectrum scarcity and fragmentation, it is difficult to provide the required resources with a single frequency band. Therefore, aggregating frequency bands, that belong to different carriers, is needed to utilize the radio resources across multiple carriers and expand the effective bandwidth delivered to user terminals, leading to inter-band non-contiguous carrier aggregation [4].

1.1 Motivation and Background

Carrier aggregation is one of the most distinct features of 4G systems including LTE-Advanced. Given the fact that LTE requires wide carrier bandwidths to utilize such as 10 and 20 MHz, CA needs to be taken into consideration when designing the system to overcome the spectrum scarcity challenges. With the CA being defined in [5], two or more component carriers (CCs) of the same or different bandwidths can be aggregated to achieve wider transmission bandwidths between the evolve node B (eNodeB) and the UE. This feature allows LTE-Advanced to meet the International Mobile Telecommunications (IMT) requirements for the fourth-generation standards defined by the International Telecommunications Union (ITU) [6]. An overview of CA framework and cases is presented in [3]. Many operators are

H. Shajaiah et al., *Resource Allocation with Carrier Aggregation in Cellular Networks*, DOI 10.1007/978-3-319-60540-1_1

willing to add the CA feature to their plans across a mixture of macro cells and small cells. This will provide capacity and performance benefits in areas where small cell coverage is available while enabling network operators to provide robust mobility management on their macro cell networks.

The non-contiguous carrier aggregation task is a challenging. The challenges are both in hardware implementation and joint optimal resource allocation. Hardware implementation challenges are in the need for multiple oscillators, multiple RF chains, more powerful signal processing, and longer battery life [7].

Increasing the utilization of the existing spectrum can significantly improve network capacity, data rates, and user experience. Some spectrum holders such as government users do not use their entire allocated spectrum in every part of their geographic boundaries most of the time. Therefore, the National Broadband Plan (NBP) and the findings of the President's Council of Advisors on Science and Technology (PCAST) spectrum study have recommended making the under-utilized federal spectrum available for secondary use [8]. Spectrum sharing enables wireless systems to use the under-utilized spectrum efficiently. Making more spectrum available can provide significant gain in mobile broadband capacity only if those resources can be aggregated efficiently with the existing commercial mobile system resources. As a result of the high demand for spectrum by commercial wireless operators, federal agencies are now willing to share their spectrum with commercial users. This has led to proposals to share spectrum allocated for federal radar operations with commercial users. The 3550–3650 MHz band, currently used for military radar operations, is identified for spectrum sharing between military radars and communication systems, according to the NTIA's 2010 Fast Track Report [9]. This band is very favorable for commercial cellular systems such as LTE-Advanced systems. Therefore, innovative methods are required to make spectrum sharing between radars and cellular systems a reality.

Beside CA capability, next-generation wireless networks need to support diverse QoS requirements of multiple applications since different applications require different application's performance. Furthermore, certain types of users may require to be given priority when allocating the network resources (i.e., such as public safety users) which needs to be taken into consideration when designing the resource allocation framework.

The public safety wide area wireless communication system is currently separate from the commercial cellular networks. Industries are willing to support both communities by providing a common technology. Release 12 of 3GPP LTE standards has enhanced LTE to support public safety requirements. Advanced standards such as LTE provide multimedia capabilities and voice and messages services at multi-megabit per second. The services that public safety networks provide such as communications for police, fire, and ambulance require systems development to meet the communication needs of emergency services. A common technical standard for commercial and public safety users provides advantages for both. The public safety systems market is much smaller than the commercial cellular market which makes it unable to attract the level of investment that goes in to commercial cellular networks and this makes a common technical standards for both the best solution. The public safety community gains access to the technical advantages provided by the commercial cellular networks whereas the commercial cellular community gains enhancement in their systems and make it more attractive to consumers. The USA has reserved spectrum in the 700 MHz band for an LTE based public safety network. The current public safety standards support medium speed data which drives the need of new technology. An efficient resource allocation framework is needed for cellular networks that support both of commercial and public safety communities and takes into consideration that users' applications should not be treated evenly for both communities.

1.2 Carrier Aggregation

1.2.1 Motivation for Developing Carrier Aggregation with a QoE approach

Historically, mobile communications researchers have focused on the important aspect of quality of service (QoS) for improving network service performance, as shown in [1, 10, 11], or recently more emphasis on quality of experience (QoE) emerges, as shown in [12, 13], for a subjective end user experience and satisfaction. With the increasing demand for mobile services [14–17], more attention to QoE is on the rise. Therefore, we can find QoS or QoE being the focus of research on various layers of the Open Systems Interconnection (OSI) model [18] and with new and creative methods being conducted. For example, QoS research results on the network layer were conducted in [19, 20], new game theory methods were conducted in [21, 22], and microeconomics utilization in [23, 24].

Various wireless standards were considered when improving QoS. For instance, QoS improvement with energy efficiency consideration was studied in [25–27] for LTE third-generation partnership project (3GPP) [28–30]. Other systems and standards were considered, e.g., QoS improvement in [31, 32] for WiMAX [33], in [34] for Universal Mobile Terrestrial System (UMTS) [35, 36], and in [37] for Mobile Broadband [38]. Moreover, application layer QoE was the focus of the research conducted in [39, 40].

Cross-layer design of OSI model was considered for QoS improvement [41, 42]. For example, router shaping and scheduling was studied in [43, 44] for Asynchronous Transfer Mode (ATM), and in [45, 46] and [47–49] for Integrated and Differentiated Services, respectively. Battery of life and QoS improvement were studied in [50–54].

For QoS and QoE improvement, resource allocation optimization problem had been considered in [22, 55] for elastic traffic, [56–58] with proportional fairness, in [59–62] with max-min fairness. The optimal solution for elastic case was considered in [63, 64] for proportional fair case and in [65, 66] for weighted fair queuing case. In [7], an approximate solution for inelastic traffic was attempted for multi-class service offering [67]. In [68, 69], authors showed the optimal solution for the inelastic traffic case. This was followed by an extension to include multiple applications per user with inelastic traffic in [40, 70–72].

The idea of using multi-carrier has been driven by the rapid data user growth and the increasing demand for resources, e.g., carrier aggregation with resource allocation [73, 74]. Given the President Council of Advisers on Science and Technology report [75], the future of resource allocation is carrier aggregation between heterogeneous spectra [76–78]. Hence, the Federal Communications Commission (FCC) recommended radar band as a secondary spectrum [79, 80] with cellular band [81, 82], and the National Telecommunications and Information Administration (NTIA) studied interference effects of radar/comm coexistence and provided useful studies on that [83–85]. Additionally, operators are facing operational challenges in terms of data capacity. The carrier aggregation feature has been added to Release 10 of the 3GPP LTE-Advance standard to allow single users to employ multiple carriers in order to achieve higher bandwidth [86]. With the increasing number of applications and their required bandwidth, smart phones now require large bandwidth allocations which makes them limited to the network resources. The peak data rates required by International Mobile Telecommunications-Advanced (IMT-Advanced) can be satisfied by LTE-Advanced as it supports wider bandwidth by using the carrier aggregation feature.

Carrier aggregation is also needed because of the fact that the current frequency spectrum is highly segmented [87]. Figure 1.1 [88] shows the current frequency allocation table for the US and how segmented the spectrum is. Fragmented spectrum can be utilized more efficiently by aggregating non-contiguous carriers.

The overall goal of carrier aggregation is to provide an enhanced QoS for mobile users throughout the cell by combining peak capacities available at different frequencies, providing more consistent QoS to users by utilizing unused capacity available at other frequencies, improving mobility, and enabling interference management.

1.2.2 Deployment Scenarios for Carrier Aggregation

Different deployment scenarios have been considered for the design of LTE-Advanced carrier aggregation [89]. Figure 1.2 shows five different deployment scenarios with two component carriers F1 and F2. The five scenarios are described below.

Scenario 1: Cells with the two carrier frequencies are collocated and overlaid in the same band. Both frequencies F1 and F2 almost have the same coverage area. Carrier aggregation enables a higher achievable data rates throughput the cell.

Scenario 2: Cells with the two carriers are collocated and overlaid in different bands. Different carriers have different coverage because higher frequency bands have larger path loss. Higher frequency bands carriers are used to improve data rates.

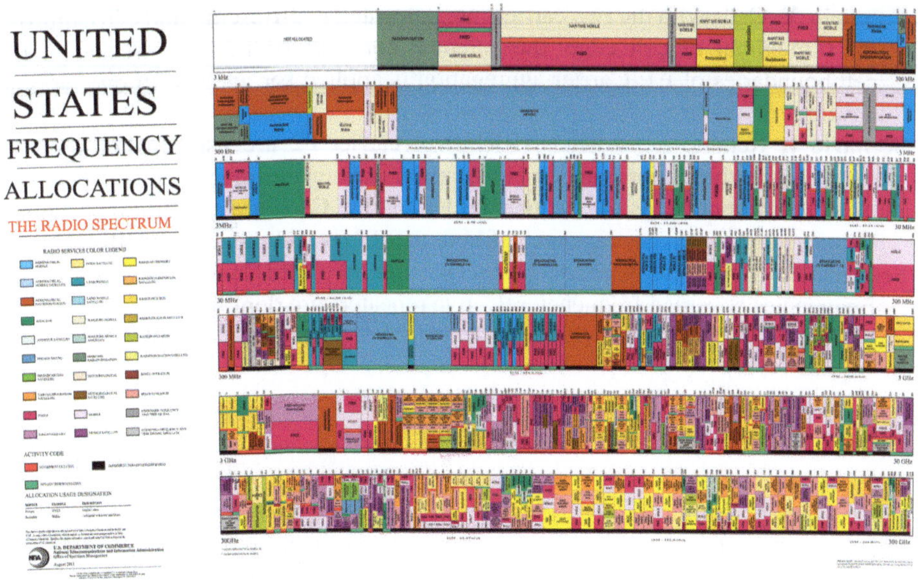

Fig. 1.1 US frequency spectrum allocation

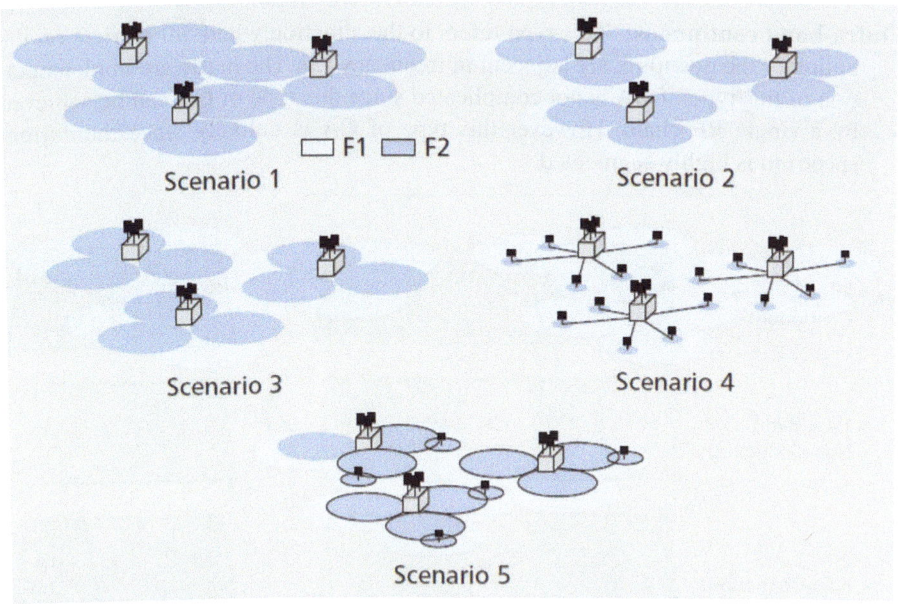

Fig. 1.2 Carrier aggregation deployment scenarios with F2 > F1 [89]

Scenario 3: Cells with the two carriers are colocated in different bands. To improve the throughput of cell edge, the antennas for cells of F2 are directed to the cell boundaries of F1. Carrier aggregation is applied for areas with overlapping coverage.

Scenario 4: Remote radio heads (RRHs) of carrier F2 are used in hot spots to improve the throughput and cells of carrier F1 are the macro cells. There are usually different bands for frequencies F1 and F2. Carrier aggregation is applied for users under the coverage area of both the RRHs and the macro cells.

Scenario 5: Similar to scenario 2 except that in order to extend one of the frequencies coverage frequency selective repeaters are deployed.

1.2.3 Types of Carrier Aggregation

Three types of carrier aggregation have been defined in 3GPP in order to meet operators spectrum scenarios. These types are intra-band contiguous, intra-band non-contiguous, and inter-band non-contiguous [3]. The uplink and downlink can be configured independently. However, the number of uplink carriers need not to exceed the number of downlink carriers. The three types of CA are illustrated in Fig. 1.3 and discussed below.

Intra-band continuous: This type refers to the situation where all carriers on the uplink or the downlink are adjacent in frequency [3]. The hardware implementation of this type of CA is not complicated since this type of CA can be achieved by a single RF chain. However this type of CA is unlikely since the current spectrum is highly segmented.

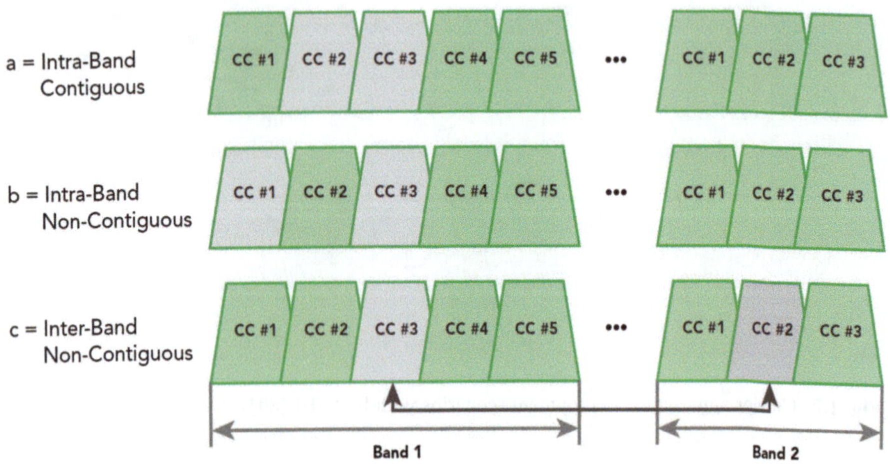

Fig. 1.3 Types of carrier aggregation in LTE-Advanced

Intra-band non-contiguous: In this type of CA, the combined carriers fall within the same band but are not adjacent in frequency [3]. This type is more realistic since the frequency bands are highly segmented. The hardware implementation of this type can simply be achieved through a single RF chain given that carriers are in the same frequency band.

Inter-band non-contiguous: In this type of CA, the two carriers are within different bands [3]. The user hardware implementation for this type is the most complex since a single RF chain has limitation in terms of a certain band of interest for practical reasons.

1.3 Previous Studies in Resource Allocation for Spectrum Sharing

There has been several works in the area of resource allocation optimization to utilize the scarce radio spectrum efficiently. Utility functions have been widely used in the formulation of different resource allocation problems. Max-min fairness [59, 90–94] and proportional fairness [57] have been used in the formulation for the allocation policy as they can achieve optimal or efficient solutions [22, 95, 96]. The authors in [63, 97, 98] have used a strictly concave utility function to represent each user's elastic traffic and proposed distributed algorithms at the sources and the links to interpret the congestion control of communication networks. Their work has only focused on elastic traffic and did not consider real-time applications as it has non-concave utility functions as shown in [99].

The authors in [100] and [101] have argued that the utility function, which represents the user application performance, is the one that needs to be shared fairly rather than the bandwidth. In this research work, we consider using resource allocation to achieve a utility proportional fairness that maximizes the user satisfaction. If a bandwidth proportional fairness is applied through a max-min bandwidth allocation, users running delay-tolerant applications receive larger utilities than users running real-time applications as real-time applications require minimum encoding rates and their utilities are equal to zero if they do not receive their minimum encoding rates.

The proportional fairness framework of Kelly introduced in [63] does not guarantee a minimum QoS for each user application. To overcome this issue, a resource allocation algorithm that uses utility proportional fairness policy is introduced in [102]. We believe that this approach is more appropriate as it respects the inelastic behavior of real-time applications. The utility proportional fairness approach in [102] gives real-time applications priority over delay-tolerant applications when allocating resources and guarantees that no user is allocated zero rate. In [102, 103] and [104], the authors have presented optimal resource allocation algorithms to allocate single carrier resources optimally among mobile users. However, their algorithms do not support multi-carrier resource allocation. To incorporate the carrier aggregation feature, we have introduced a multi-stage resource allocation using carrier aggregation in [73]. In [105] and [74], we present resource allocation with users discrimination algorithms to allocate the eNodeB resources optimally among mobile users with elastic and inelastic traffic. In [106], the authors have presented a radio resource block allocation optimization problem using a utility proportional fairness approach. The authors in [40] have presented an application-aware resource block scheduling approach for elastic and inelastic adaptive real-time traffic where users are assigned to resource blocks.

On the other hand, resource allocation for single cell multi-carrier systems has been given extensive attention in recent years [107–109]. In [110–113], the authors have represented this challenge in optimization problems. Their objective is to maximize the overall cell throughput with some constraints such as fairness and transmission power. However, transforming the problem into a utility maximization framework can achieve better users satisfaction rather than better system-centric throughput. Also, in practical systems, the challenge is to perform multi-carrier radio resource allocation for multiple cells. The authors in [114, 115] suggested using a distributed resource allocation rather than a centralized one to reduce the implementation complexity. In [116], the authors propose a collaborative scheme in a multiple base stations (BSs) environment, where each user is served by the BS that has the best channel gain with that user. The authors in [117] have addressed the problem of spectrum resource allocation in carrier aggregation based LTE-Advanced systems, with the consideration of UEs' MIMO capability and the modulation and coding schemes (MCSs) selection.

Most of the previous research work have focused on finding resource allocation approaches for intra-system and intra-operator of a single network operator. However, current research on resource allocation is for more complex network topologies [118, 119]. Carrier aggregation in networks that involve multiple network operators in HetNets need to be further investigated. In [120], the authors have analyzed the performance of their proposed carrier aggregation framework that combines a statically assigned spectrum with spectrum resources from a shared spectrum pool. A tractable multi-band multi-tier CA models for HetNets are proposed in [121]. Two models are considered: multi-flow CA and single-flow CA, each UE performs cell selection based on the reference signal's maximum received power. A major concern about deploying small cells is their small coverage areas and low transmit power. The authors in [122, 123] have addressed this issue and suggested biasing to allow small cells to expand their coverage areas.

In the past, wireless systems were able to share government bands by operating on a low power to prevent the interference with the incumbent systems such as wireless local area network (WLAN) in the 5.25–5.35 and 5.47–5.725 GHz radar bands [80]. Small cells operating in a low power have been proposed recently to operate in the 3.5 GHz radar band [124].

To mitigate radar interference to LTE-Advanced systems, a spatial approach for spectrum sharing between a MIMO radar and LTE cellular system with N_{BS} base stations was proposed in [125, 126]. Radar signals are manipulated such that they are not a source of interference to the LTE-Advanced BSs. Because there exist many interference channels between the two systems, the interference channel with the maximum null-space dimension is chosen based on the algorithm proposed by the authors, the radar signal is then projected onto the null space of that interference channel to mitigate interference to the LTE-Advanced BS. This spatial approach results in small degradation in the radar performance [96].

In [96], the authors proposed a technique to project radar waveforms onto the null space of an interference channel matrix between the radar and the communication system. In their proposed approach, the cognitive radar is assumed to have full knowledge of the interference channel and modifies its signal vectors in a way such that they are in the null space of the channel matrix. In order to avoid interference to the communication system, a radar signal projection onto the null space of interference channel between radar and communication systems is presented in [126, 127]. In [128], a novel signal processing approach is developed for coherent MIMO radar to minimize the arbitrary interferences generated by wireless systems from any direction while operating at the same frequency using cognitive radio technology.

Some researchers introduced carrier aggregation scenarios using non-convex optimization methods in [129–133]. Further inclusion of radar band in particular as a secondary band was provided in [134–136] for the radar/comm coexistence problem [137–140].

As 4G wireless mobile systems including LTE and LTE-Advanced continue to evolve, higher data rates and improved QoS, even for cell-edge users, are promised to be guaranteed for end users. The capacity promised by MIMO systems may not be fully realizable without having a sufficient control of inter-cell interference which limits throughput for cell-edge users [141]. In order to mitigate inter-cell interference, three major frequency reuse patterns can be used: hard frequency reuse, fractional frequency reuse (FFR), and soft frequency reuse. Hard frequency reuse divides the system bandwidth into a number of sub-bands according to certain reuse factor such that neighboring cells transmit on different sub-bands. FFR divides the system bandwidth into an inner and an outer part. The inner part is only allocated to the near users with reduced power while applying a frequency reuse factor of 1 such that the inner part is reused by all other BSs. On the other hand, the outer part of bandwidth is allocated to far users (cell-edge users) with a frequency reuse factor greater than one. Soft frequency reuse allows the overall bandwidth to be shared by all BSs with a reuse factor of 1 while the BSs are restricted to certain power bound for the transmission on each subcarrier. Hard frequency reuse suffers from a reduced spectral efficiency whereas soft frequency reuse [142, 143] has full spectral efficiency but requires centralized coordination of resource allocation which becomes impractical for a large number of BSs. Unless otherwise specified, we consider using FFR as it compromises between hard and soft frequency reuse and therefore will be a proficient option for future wireless systems. However, we do not intend to address inter-cell interference throughout the book chapters.

1.4 Overview

An overview of the contributions to this book is given for each topic as follows:

Multi-Stage Resource Allocation with Carrier Aggregation:

- We formulate the resource allocation with CA problem into a convex optimization framework and develop a utility proportional fairness RA with CA optimization problem that gives priority to real-time applications due to the nature of the sigmoidal-like utility functions [144, 145].
- We develop a distributed multi-stage resource allocation algorithm to solve the optimization problem and allocate the eNodeBs' resources optimally among users, with a minimum QoS guaranteed for each user, and present its corresponding simulation results.
- We develop a price selective centralized resource allocation with CA scheme to allocate multiple carriers resources optimally among users located under one carrier or multiple carriers coverage area and present the corresponding algorithm that allows each UE to choose its primary carrier and the order of its secondary carriers based on the price offered by all in range carriers.
- We present simulation results for the performance of the proposed price selective centralized algorithm and show how it converges to the optimal rates whether the eNodeBs' available resources are abundant or scarce.

Resource Allocation with User Discrimination:

- We develop a spectrum sharing scheme for public safety and commercial users running elastic or inelastic traffic and formulate a resource allocation optimization problem to allocate the eNodeB resources optimally among public safety and commercial users. In addition, we present a resource allocation algorithm to allocate an optimal rate to each UE with a priority given to public safety users. Within the same group of users, a priority is given to real-time applications presented by sigmoidal-like utility functions.
- We develop a resource allocation with user discrimination framework to allocate a single carrier resources optimally among different types of users running multiple applications.
- We propose a two-stage rate allocation method for the single carrier RA with user discrimination optimization problem and present its corresponding algorithms. First, the eNodeB and the UE collaborate to allocate an optimal rate to each UE. Each UE then allocates its assigned rate optimally among its applications.
- We develop a multi-stage resource allocation with user discrimination optimization problem to allocate multi-carrier resources optimally among different classes of users. In addition, we prove that the resource allocation optimization problem is convex and therefore the global optimal solution is tractable.
- We present a resource allocation algorithm to solve the multi-stage RA with user discrimination optimization problem and allocate each user an aggregated final rate from its in range carriers. We present simulation results for the performance of the proposed algorithm.

Resource Allocation with Carrier Aggregation for Commercial Use of 3.5 GHz Spectrum:

- We develop a spectrum sharing approach, based on multi-stage resource allocation with CA, for sharing the Federal under-utilized 3.5 GHz spectrum with commercial users and present its corresponding simulation results.

Resource Allocation for Spectrum Sharing Between Radar and Communication Systems:

- We present a spectrum sharing scenario between a MIMO radar and an LTE system with multiple base stations and propose a channel-selection algorithm to select the best channel for radar's signal projection that maintains a minimum degradation in the radar performance while causing no interference to the LTE BS. We also present a null-space projection (NSP) algorithm that performs the null-space computation.
- We present a multi-stage RA with CA algorithm for the proposed spectrum sharing approach to allocate both of the radar and the LTE-Advanced carriers' resources optimally among users running real-time or delay-tolerant applications. We show through simulation results that the proposed algorithm is a robust algorithm that converges to the optimal rates for high available resources and scarce resources cases.

Resource Block Scheduling with Carrier Aggregation Based on Utility Proportional Fairness:

- We develop a framework for the problem of utility proportional fairness RB scheduling with CA for multi-carrier cellular networks.
- We prove that the proposed resource scheduling policy, that is based on CA, exists and that the optimal solution is tractable. We show through simulation results the performance of the proposed resource scheduling with CA approach and compare it with other resource scheduling policies.

1.5 Book Organization

The rest of this book is organized as follows. Chapter 2 discusses the users applications utility functions used and their properties. It also presents the general form of the resource allocation optimization problem that is used throughout the book. Chapter 3 presents a multi-stage distributed and centralized resource allocation with CA framework. Chapter 4 develops a spectrum sharing architecture between commercial and public safety cellular systems and provides a resource allocation with user discrimination framework for multi-carrier cellular networks. Chapter 5 provides resource allocation with CA framework and algorithms for the spectrum sharing scenario that involves commercial use of the 3.5 GHz spectrum. Chapter 6 presents a spectrum sharing approach between radar and communication systems and provides a resource allocation with CA approach for an LTE-Advanced Cellular System Sharing Spectrum with S-band Radar. Chapter 7 presents a utility proportional RB scheduling with CA approach and compares the proposed scheduling policy with other resource scheduling policies. Chapter 8 summarizes the contributions presented in this book and points out some future extensions of this work.

References

1. H. Ekstrom, QoS control in the 3GPP evolved packet system. IEEE Commun. Mag. **47**, 76–83 (2009)
2. Cisco, Visual networking index. White paper at Cisco.com, February 2014
3. M. Iwamura, K. Etemad, M.-H. Fong, R. Nory, R. Love, Carrier aggregation framework in 3GPP LTE-advanced [WiMAX/LTE Update]. IEEE Commun. Mag. **48**(8), 60–67 (2010)
4. Y. Wang, K.I. Pedersen, T.B. Sørensen, P.E. Mogensen, Utility maximization in LTE-advanced systems with carrier aggregation, in *VTC Spring* (2011), pp. 1–5
5. G. RP-091440, Work item description: carrier aggregation for LTE, December 2009
6. S. Parkvall, A. Furuskar, E. Dahlman, Evolution of LTE toward IMT-advanced. IEEE Commun. Mag. **49**, 84–91 (2011)
7. R.L. Kurrle, Resource allocation for smart phones in 4G LTE advanced carrier aggregation, November 2012
8. PCAST, Final PCAST spectrum report , July 2012
9. National Telecommunications and Information Administration (NTIA), An assessment of the near-term viability of accommodating wireless broadband systems in the 1675–1710 MHz, 1755–1780 MHz, 3500–3650 MHz, 4200–4220 MHz, and 4380–4400 MHz bands (Fast Track Report). Online, October 2010
10. H. Ekstrom, A. Furuskar, J. Karlsson, M. Meyer, S. Parkvall, J. Torsner, M. Wahlqvist, Technical solutions for the 3G long-term evolution. IEEE Commun. Mag. **44**, 38–45 (2006)
11. M. Ghorbanzadeh, A. Abdelhadi, C. Clancy, Quality of service in communication systems, in *Cellular Communications Systems in Congested Environments* (Springer, Berlin, 2017), pp. 1–20
12. S. Qaiyum, I.A. Aziz, J.B. Jaafar, Analysis of big data and quality-of-experience in high-density wireless network, in *2016 3rd International Conference on Computer and Information Sciences (ICCOINS)*, August 2016, pp. 287–292
13. A. Ghosh, R. Ratasuk, Essentials of LTE and LTE-A (2011)
14. I. Research, Mobile VoIP subscribers will near 410 million by 2015; VoLTE still a long way off (2010)
15. Nokia Solutions and Networks, Enhance mobile networks to deliver 1000 times more capacity by 2020 (2013)
16. G. Intelligence, Smartphone users spending more ‘face time’ on apps than voice calls or web browsing (2011)
17. Nokia Siemens Networks, Understanding Smartphone Behavior in the Network (2011)
18. W. Stallings, Data and computer communications, in *William Stallings Books on Computer and Data Communications* (Pearson Prentice Hall, Pearson Education, Inc., Upper Saddle River, 2013)
19. A. Larmo, M. Lindstrom, M. Meyer, G. Pelletier, J. Torsner, H. Wiemann, The LTE link-layer design (2009)
20. C. Ciochina, H. Sari, A review of OFDMA and single-carrier FDMA, in *Wireless Conference (EW)* (2010)

21. S. Ali, M. Zeeshan, A delay-scheduler coupled game theoretic resource allocation scheme for LTE networks, in *Frontiers of Information Technology (FIT)* (2011)
22. D. Fudenberg, J. Tirole, *Nash Equilibrium: Multiple Nash Equilibria, Focal Points, and Pareto Optimality* (MIT Press, New York, 1991)
23. P. Ranjan, K. Sokol, H. Pan, Settling for less - a QoS compromise mechanism for opportunistic mobile networks, in *SIGMETRICS Performance Evaluation* (2011)
24. R. Johari, J. Tsitsiklis, Parameterized supply function bidding: equilibrium and efficiency (2011)
25. L.B. Le, E. Hossain, D. Niyato, D.I. Kim, Mobility-aware admission control with QoS guarantees in OFDMA Femtocell networks, in *2013 IEEE International Conference on Communications (ICC)*, June 2013, pp. 2217–2222
26. L.B. Le, D. Niyato, E. Hossain, D.I. Kim, D.T. Hoang, QoS-aware and energy-efficient resource management in OFDMA Femtocells. IEEE Trans. Wirel. Commun. **12**, 180–194 (2013)
27. L. Chung, Energy efficiency of QoS routing in multi-hop wireless networks, in *IEEE International Conference on Electro/Information Technology (EIT)* (2010)
28. G.T. V9.0.0, Further advancements for E-UTRA physical layer aspects (2012)
29. 3GPP Technical Report 36.211. Physical channels and modulation (2007). www.3gpp.org
30. 3GPP Technical Report 36.213, Physical layer procedures (2011). www.3gpp.org
31. M. Alasti, B. Neekzad, J. Hui, R. Vannithamby, Quality of service in WiMAX and LTE networks [Topics in Wireless Communications] (2010)
32. D. Niyato, E. Hossain, Wireless broadband access: WiMAX and beyond - integration of WiMAX and WiFi: optimal pricing for bandwidth sharing. IEEE Commun. Mag. **45**, 140–146 (2007)
33. J. Andrews, A. Ghosh, R. Muhamed, Fundamentals of WiMAX: understanding broadband wireless networking (2007)
34. F. Li, Quality of service, traffic conditioning, and resource management in universal mobile telecommunication system (UMTS) (2003)
35. European Telecommunications Standards Institute, UMTS; LTE; UTRA; E-UTRA;EPC; UE conformance specification for UE positioning; Part 1: Conformance test specification (2012)
36. European Telecommunications Standards Institute, UMTS; UTRA; General description; Stage 2 (2016)
37. IXIACOM, Quality of service (QoS) and policy management in mobile data networks (2010)
38. FCC (2010), Unlicensed operations in the TV broadcast bands, second memorandum opinion and order. Technical Report, FCC 10–174 (2010)
39. T. Erpek, A. Abdelhadi, T.C. Clancy, Application-aware resource block and power allocation for LTE, in *2016 Annual IEEE Systems Conference (SysCon)*, April 2016, pp. 1–5
40. T. Erpek, A. Abdelhadi, C. Clancy, An optimal application-aware resource block scheduling in LTE, in *IEEE International Conference on Computing, Networking and Communications (ICNC) Workshop CCS)* (2015)
41. C. Dovrolis, D. Stiliadis, P. Ramanathan, Proportional differentiated services: delay differentiation and packet scheduling (2002)
42. A. Sali, A. Widiawan, S. Thilakawardana, R. Tafazolli, B. Evans, Cross-layer design approach for multicast scheduling over satellite networks, in *2nd International Symposium on Wireless Communication Systems, 2005* (2005)
43. E. Lutz, D. Cygan, M. Dippold, F. Dolainsky, W. Papke, The land mobile satellite communication channel-recording, statistics, and channel model (1991)
44. H. Perros, K. Elsayed, Call admission control schemes: a review (1994)
45. R. Braden, Integrated services in the internet architecture: an overview (1994)
46. R. Braden, Resource reservation protocol (RSVP) - Version 1 functional specification (1997)
47. S. Blake, An architecture for differentiated services (1998)
48. K. Nichols, A two-bit differentiated services architecture for the internet (1999)
49. K. Nahrstedt, The QoS broker (1995)

50. I. Jung, I. Jo, Y. Yu, H. Eom, H. Yeom, Enhancing QoS and energy efficiency of realtime network application on smartphone using cloud computing, in *IEEE Asia-Pacific Services Computing Conference (APSCC)* (2011)
51. Tellabs, Quality of service in the wireless Backhaul (2012)
52. N. Ahmed, H. Yan, Access control for MPEG video applications using neural network and simulated annealing. Math. Probl. Eng. **2004**(3), 291–234 (2004)
53. J. Tournier, J. Babau, V. Olive, Qinna, a component-based QoS architecture, in *Proceedings of the 8th International Conference on Component-Based Software Engineering* (2005)
54. G. Gorbil, I. Korpeoglu, Supporting QoS traffic at the network layer in multi-hop wireless mobile networks, in *Wireless Communications and Mobile Computing Conference (IWCMC)* (2011)
55. M. Ghorbanzadeh, A. Abdelhadi, C. Clancy, Utility functions and radio resource allocation, in *Cellular Communications Systems in Congested Environments* (Springer, Berlin, 2017), pp. 21–36
56. H. Kushner, P. Whiting, Convergence of proportional-fair sharing algorithms under general conditions (2004)
57. M. Andrews, K. Kumaran, K. Ramanan, A. Stolyar, P. Whiting, R. Vijayakumar, Providing quality of service over a shared wireless link. IEEE Commun. Mag. **39**(2), 150–154 (2001)
58. G. Tychogiorgos, A. Gkelias, K.K. Leung, Utility-proportional fairness in wireless networks, in *PIMRC* (IEEE, New York, 2012), pp. 839–844
59. M. Li, Z. Chen, Y. Tan, A maxmin resource allocation approach for scalable video delivery over multiuser MIMO-OFDM systems, in *IEEE International Symposium on Circuits and Systems (ISCAS)* (2011)
60. R. Prabhu, B. Daneshrad, An energy-efficient water-filling algorithm for OFDM systems, in *IEEE International Conference on Communications (ICC)* (2010)
61. T. Harks, Utility proportional fair bandwidth allocation: an optimization oriented approach, in *QoS-IP* (2005), pp. 61–74
62. T. Nandagopal, T. Kim, X. Gao, V. Bharghavan, Achieving MAC layer fairness in wireless packet networks, in *Proceedings of the 6th Annual International Conference on Mobile Computing and Networking (Mobicom)* (2000)
63. F. Kelly, A. Maulloo, D. Tan, Rate control in communication networks: shadow prices, proportional fairness and stability. J. Oper. Res. Soc. **49**, 237–252 (1998)
64. S.H. Low, D.E. Lapsley, Optimization flow control, I: basic algorithm and convergence. IEEE/ACM Trans. Netw. **7**(6), 861–874 (1999)
65. A. Parekh, R. Gallager, A generalized processor sharing approach to flow control in integrated services networks: the single-node case (1993)
66. A. Demers, S. Keshav, S. Shenker, Analysis and simulation of a fair queueing algorithm (1989)
67. J. Lee, R. Mazumdar, N. Shroff, Non-convex optimization and rate control for multi-class services in the internet (2005)
68. A. Abdelhadi, A. Khawar, T.C. Clancy, Optimal downlink power allocation in cellular networks. Phys. Commun. **17**, 1–14 (2015)
69. A. Abdelhadi, T.C. Clancy, Optimal context-aware resource allocation in cellular networks, in *2016 International Conference on Computing, Networking and Communications (ICNC)*, February 2016, pp. 1–5
70. M. Ghorbanzadeh, A. Abdelhadi, C. Clancy, Resource allocation architectures traffic and sensitivity analysis, in *Cellular Communications Systems in Congested Environments* (Springer, Berlin, 2017), pp. 93–116
71. M. Ghorbanzadeh, A. Abdelhadi, C. Clancy, A utility proportional fairness approach for resource block allocation in cellular networks, in *IEEE International Conference on Computing, Networking and Communications (ICNC)* (2015)
72. M. Ghorbanzadeh, A. Abdelhadi, C. Clancy, Radio resource block allocation, in *Cellular Communications Systems in Congested Environments* (Springer, Berlin, 2017), pp. 117–146

73. H. Shajaiah, A. Abdelhadi, C. Clancy, Utility proportional fairness resource allocation with carrier aggregation in 4g-LTE, in *IEEE Military Communications Conference (MILCOM)* (2013)
74. H. Shajaiah, A. Abdelhadi, C. Clancy, Multi-application resource allocation with users discrimination in cellular networks, in *IEEE international Symposium on Personal, Indoor and Mobile Radio Communications (PIMRC)* (2014)
75. Executive office of the President, Presidents Council of Advisors on Science and Technology, Realizing the full potential of government-held spectrum to spur economic growth (2012)
76. H. Shajaiah, A. Abdelhadi, C. Clancy, A price selective centralized algorithm for resource allocation with carrier aggregation in LTE cellular networks, in *2015 IEEE Wireless Communications and Networking Conference (WCNC)*, March 2015, pp. 813–818
77. H. Shajaiah, A. Abdelhadi, C. Clancy, Spectrum sharing approach between radar and communication systems and its impact on radar's detectable target parameters, in *2015 IEEE 81st Vehicular Technology Conference (VTC Spring)* (May) 2015, pp. 1–6
78. S. Wilson, T. Fischetto, Coastline population trends in the united states: 1960 to 2008. U.S. Department of Commerce (2010)
79. M. Richards, J. Scheer, W. Holm, Principles of modern radar (2010)
80. Federal Communications Commission (FCC), In the matter of revision of parts 2 and 15 of the commission's rules to permit unlicensed national information infrastructure (U-NII) devices in the 5 GHz band. MO&O, ET Docket No. 03–122, June 2006
81. Federal Communications Commission, Proposal to create a citizen's broadband service in the 3550–3650 MHz band (2012)
82. Federal Communications Commission (FCC), Connecting America: the national broadband plan. Online (2010)
83. NTIA, An assessment of the near-term viability of accommodating wireless broadband systems in the 1675–1710 MHz, 1755–1780 MHz, 3500–3650 MHz, 4200–4220 MHz and 4380–4400 MHz bands (2010)
84. National Telecommunications and Information Administration (NTIA), Analysis and resolution of RF interference to radars operating in the band 2700–2900 MHz from broadband communication transmitters. Online, October 2012
85. M. Cotton, R. Dalke, Spectrum occupancy measurements of the 3550–3650 megahertz maritime radar band near San Diego, California (2014)
86. G. Yuan, X. Zhang, W. Wang, Y. Yang, Carrier aggregation for LTE-advanced mobile communication systems. IEEE Commun. Mag. **48**, 88–93 (2010)
87. Z. Shen, A. Papasakellariou, J. Montojo, D. Gerstenberger, F. Xu, Overview of 3GPP LTE-advanced carrier aggregation for 4G wireless communications. IEEE Commun. Mag. **50**, 122–130 (2012)
88. *Frequency Spectrum Wall Chart* (Commerce Dept., National Telecommunications and Information Administration, Office of Spectrum Management, 2016)
89. 3GPP, Evolved universal terrestrial radio access (E-UTRA) and evolved universal terrestrial radio access network (E-UTRAN); overall description; stage 2 (Release 10). TSG RAN. TS 36.300 v10.3.0 (2011)
90. A. Khawar, A. Abdelhadi, T.C. Clancy, Channel modeling between seaborne MIMO radar and MIMO cellular system. CoRR, abs/1504.04325 (2015)
91. R.M.J.-W. Lee, N.B. Mandayam, A utility based approach for multi-class wireless systems. IEEE/ACM Trans. Netw. **13**, 854–867 (2015)
92. S. Ahmed, J.S. Thompson, B. Mulgrew, Y. Petillot, Constant envelope waveform design for MIMO radar, in *International Conference on Acoustics, Speech, and Signal Processing* (2010), pp. 4066–4069
93. A. Khawar, A. Abdelhadi, T.C. Clancy, 3D channel modeling between seaborne MIMO radar and MIMO cellular system. CoRR, abs/1504.04333 (2015)
94. A. Khawar, A. Abdelhadi, T.C. Clancy, Coexistence analysis between radar and cellular system in LOS channel. IEEE Antennas Wirel. Propag. Lett. **15**, 972–975 (2016)

95. C. Shahriar, A. Abdelhadi, T.C. Clancy, Overlapped-MIMO radar waveform design for coexistence with communication systems, in *2015 IEEE Wireless Communications and Networking Conference (WCNC)*, March 2015, pp. 223–228
96. A. Khawar, A. Abdel-Hadi, T. Clancy, R. McGwier, Beampattern analysis for MIMO radar and telecommunication system coexistence, in *2014 International Conference on Computing, Networking and Communications (ICNC)*, February 2014, pp. 534–539
97. S. Low, F. Paganini, J. Doyle, Internet congestion control. IEEE Control. Syst. **22**, 28–43 (2002)
98. J. Mo, J. Walrand, Fair end-to-end window-based congestion control. IEEE/ACM Trans. Netw. **8**, 556–567 (2000)
99. S. Shenker, Fundamental design issues for the future internet. IEEE J. Sel. Areas Commun. **13**, 1176–1188 (1995)
100. Z. Cao, E. Zegura, Utility max-min: an application-oriented bandwidth allocation scheme, in *Eighteenth Annual Joint Conference of the IEEE Computer and Communications Societies, Proceedings. IEEE INFOCOM '99* vol. 2 (1999), pp. 793–801
101. S. Sarkar, L. Tassiulas, Fair allocation of utilities in multirate multicast networks: a framework for unifying diverse fairness objectives. IEEE Trans. Autom. Control **47**, 931–944 (2002)
102. A. Abdelhadi, C. Clancy, A utility proportional fairness approach for resource allocation in 4G-LTE, in *IEEE International Conference on Computing, Networking, and Communications (ICNC), CNC Workshop* (2014)
103. A. Abdelhadi, C. Clancy, A robust optimal rate allocation algorithm and pricing policy for hybrid traffic in 4G-LTE, in *IEEE international Symposium on Personal, Indoor, and Mobile Radio Communications (PIMRC)* (2013)
104. A. Abdelhadi, C. Clancy, J. Mitola, A resource allocation algorithm for multi-application users in 4G-LTE, in *MobiCom Workshop* (2013)
105. H. Shajaiah, A. Abdelhadi, C. Clancy, Spectrum sharing between public safety and commercial users in 4G-LTE, in *IEEE International Conference on Computing, Networking and Communications (ICNC)* (2014)
106. M. Ghorbanzadeh, A. Abdelhadi, C. Clancy, A utility proportional fairness radio resource block allocation in cellular networks. arXiv:1406.2630v1 (2014)
107. M. Awad, V. Mahinthan, M. Mehrjoo, X. Shen, J.W. Mark, A dual-decomposition-based resource allocation for OFDMA networks with imperfect CSI. IEEE Trans. Veh. Technol. **59**, 2394–2403 (2010)
108. M. Mehrjoo, S. Moazeni, X.S. Shen, Resource allocation in OFDMA networks based on interior point methods. Wirel. Commun. Mob. Comput. **10**(11), 1493–1508 (2010)
109. P. Tejera, W. Utschick, J. Nossek, G. Bauch, Rate balancing in multiuser MIMO OFDM systems. IEEE Trans. Commun. **57**, 1370–1380 (2009)
110. L. Xu, X. Shen, J.W. Mark, Fair resource allocation with guaranteed statistical QoS for multimedia traffic in wideband CDMA cellular network. IEEE Trans. Mob. Comput. **4**, 166–177 (2005)
111. M. Mehrjoo, M. Awad, M. Dianati, X. Shen, Design of fair weights for heterogeneous traffic scheduling in multichannel wireless networks. IEEE Trans. Commun. **58**, 2892–2902 (2010)
112. R. Madan, S. Boyd, S. Lall, Fast algorithms for resource allocation in wireless cellular networks. IEEE/ACM Trans. Netw. **18**, 973–984 (2010)
113. Y.-B. Lin, T.-H. Chiu, Y.-T. Su, Optimal and near-optimal resource allocation algorithms for OFDMA networks. IEEE Trans. Wirel. Commun. **8**, 4066–4077 (2009)
114. G. Li, H. Liu, Downlink dynamic resource allocation for multi-cell OFDMA system, in *Conference Record of the Thirty-Seventh Asilomar Conference on Signals, Systems and Computers, 2004*, vol. 1 (2003), pp. 517–521
115. S. Cicalo, V. Tralli, A. Perez-Neira, Centralized vs distributed resource allocation in multi-cell OFDMA systems, in *2011 IEEE 73rd Vehicular Technology Conference (VTC Spring)*, May 2011, pp. 1–6
116. M. Dianati, X. Shen, K. Naik, Cooperative fair scheduling for the downlink of CDMA cellular networks. IEEE Trans. Veh. Technol. **56**, 1749–1760 (2007)

117. P.-L. Tsai, K.-J. Lin, W.-T. Chen, Downlink radio resource allocation with carrier aggregation in MIMO LTE-advanced systems, in *2014 IEEE International Conference on Communications (ICC)*, June 2014, pp. 2332–2337
118. T. Yang, L. Zhang, L. Yang, Cognitive-based distributed interference management for home-eNB systems with single or multiple antennas, in *2010 IEEE 21st International Symposium on Personal Indoor and Mobile Radio Communications (PIMRC)*, September 2010, pp. 1260–1264
119. A. Attar, V. Krishnamurthy, O. Gharehshiran, Interference management using cognitive base-stations for UMTS LTE. IEEE Commun. Mag. **49**, 152–159 (2011)
120. J. McMenamy, I. Macaluso, N. Marchetti, L. Doyle, A methodology to help operators share the spectrum through an enhanced form of carrier aggregation, in *2014 IEEE International Symposium on Dynamic Spectrum Access Networks (DYSPAN)*, April 2014, pp. 334–345
121. X. Lin, J. Andrews, R. Ratasuk, B. Mondal, A. Ghosh, Carrier aggregation in heterogeneous cellular networks, in *2013 IEEE International Conference on Communications (ICC)*, June 2013, pp. 5199–5203
122. A. Ghosh, N. Mangalvedhe, R. Ratasuk, B. Mondal, M. Cudak, E. Visotsky, T. Thomas, J. Andrews, P. Xia, H. Jo, H. Dhillon, T. Novlan, Heterogeneous cellular networks: from theory to practice. IEEE Commun. Mag. **50**, 54–64, (2012)
123. A. Damnjanovic, J. Montojo, Y. Wei, T. Ji, T. Luo, M. Vajapeyam, T. Yoo, O. Song, D. Malladi, A survey on 3GPP heterogeneous networks. IEEE Wirel. Commun. **18**, 10–21 (2011)
124. Federal Communications Commission (FCC), FCC proposes innovative small cell use in 3.5 GHz band. Online: http://www.fcc.gov/document/fcc-proposes-innovative-small-cell-use-35-ghz-band, 12 December 2012
125. A. Khawar, A. Abdel-Hadi, C. Clancy, Spectrum sharing between S-band radar and LTE cellular system: a spatial approach (2012) IEEE DySPAN-SSPARC Workshop
126. A. Khawar, A. Abdelhadi, T.C. Clancy, *Spectrum Sharing Between Radars and Communication Systems: A MATLAB Based Approach.* (Springer International Publishing, Cham, 2017)
127. A. Khawar, A. Abdelhadi, T.C. Clancy, MIMO radar waveform design for spectrum sharing with cellular systems: a MATLAB based approach. (Springer International Publishing, Cham, 2016)
128. H. Deng, B. Himed, Interference mitigation processing for spectrum-sharing between radar and wireless communications systems. IEEE Trans. Aerosp. Electron. Syst. **49**, 1911–1919 (2013)
129. G. Tychogiorgos, A. Gkelias, K.K. Leung, A new distributed optimization framework for hybrid ad-hoc networks, in *GLOBECOM Workshops* (2011), pp. 293–297
130. G. Tychogiorgos, A. Gkelias, K.K. Leung, Towards a fair non-convex resource allocation in wireless networks, in *PIMRC* (2011), pp. 36–40
131. T. Jiang, L. Song, Y. Zhang, *Orthogonal Frequency Division Multiple Access Fundamentals and Applications* (Auerbach Publications, Boca Raton, 2010)
132. G. Yuan, X. Zhang, W. Wang, Y. Yang, Carrier aggregation for LTE-advanced mobile communication systems. IEEE Commun. Mag. **48**, 88–93 (2010)
133. Y. Wang, K. Pedersen, P. Mogensen, T. Sorensen, Resource allocation considerations for multi-carrier LTE-advanced systems operating in backward compatible mode, in *2009 IEEE 20th International Symposium on Personal, Indoor and Mobile Radio Communications* (2009), pp. 370–374
134. M. Ghorbanzadeh, A. Abdelhadi, C. Clancy, A utility proportional fairness bandwidth allocation in radar-coexistent cellular networks, in *Military Communications Conference (MILCOM)* (2014)
135. A. Abdelhadi, T.C. Clancy, Network MIMO with partial cooperation between radar and cellular systems, in *2016 International Conference on Computing, Networking and Communications (ICNC)*, February 2016, pp. 1–5
136. M. Ghorbanzadeh, A. Abdelhadi, C. Clancy, Spectrum-shared resource allocation, in *Cellular Communications Systems in Congested Environments* (Springer, Berlin, 2017), pp. 147–178

137. A. Lackpour, M. Luddy, J. Winters, Overview of interference mitigation techniques between WiMAX networks and ground based radar (2011)
138. F. Sanders, J. Carrol, G. Sanders, R. Sole, Effects of radar interference on LTE base station receiver performance (2013)
139. M.P. Fitz, T.R. Halford, I. Hossain, S.W. Enserink, Towards simultaneous radar and spectral sensing, in *IEEE International Symposium on Dynamic Spectrum Access Networks (DYSPAN)*, April 2014, pp. 15–19
140. Z. Khan, J.J. Lehtomaki, R. Vuohtoniemi, E. Hossain, L.A. Dasilva, On opportunistic spectrum access in radar bands: lessons learned from measurement of weather radar signals. IEEE Wirel. Commun. **23**, 40–48 (2016)
141. H. Huang, M. Trivellato, A. Hottinen, M. Shafi, P. Smith, R. Valenzuela, Increasing downlink cellular throughput with limited network MIMO coordination. IEEE Trans. Wirel. Commun. **8**, 2983–2989 (2009)
142. A. Gjendemsjo, D. Gesbert, G. Oien, S. Kiani, Optimal power allocation and scheduling for two-cell capacity maximization, in *2006 4th International Symposium on Modeling and Optimization in Mobile, Ad Hoc and Wireless Networks*, April 2006, pp. 1–6
143. S. Das, H. Viswanathan, G. Rittenhouse, Dynamic load balancing through coordinated scheduling in packet data systems, in *INFOCOM 2003. Twenty-Second Annual Joint Conference of the IEEE Computer and Communications. IEEE Societies*, vol. 1, March 2003, pp. 786–796
144. Y. Wang, A. Abdelhadi, T.C. Clancy, Optimal power allocation for LTE users with different modulations, in *2016 Annual IEEE Systems Conference (SysCon)*, April 2016, pp. 1–5
145. M. Ghorbanzadeh, A. Abdelhadi, C. Clancy, Distributed resource allocation, in *Cellular Communications Systems in Congested Environments* (Springer, Berlin, 2017), pp. 61–91

Chapter 2
Utility Functions and Resource Allocation for Spectrum Sharing

2.1 User Applications Utility Functions

The user satisfaction with the provided service can be expressed using utility functions that represent the degree of satisfaction of the user function of the rate allocated by the cellular network [1–9]. We assume that the applications utility functions $U(r)$ are strictly concave or sigmoidal-like functions [10–16].

These applications utility functions have the following properties:

- $U(0) = 0$ and $U(r)$ is an increasing function of r.
- $U(r)$ is twice continuously differentiable in r and bounded above.

We use the normalized sigmoidal-like utility function, same as the one presented in [5, 17, 18], that is

$$U(r) = c\Big(\frac{1}{1 + e^{-a(r-b)}} - d\Big), \tag{2.1}$$

where $c = \frac{1+e^{ab}}{e^{ab}}$ and $d = \frac{1}{1+e^{ab}}$ so it satisfies $U(0) = 0$ and $U(\infty) = 1$ [16]. The normalized sigmoidal-like function has an inflection point at $r^{\text{inf}} = b$. In addition, we use the normalized logarithmic utility function, used in [1, 2, 19–21], that can be expressed as

$$U(r) = \frac{\log(1 + kr)}{\log(1 + kr_{\text{max}})}, \tag{2.2}$$

where r_{max} gives 100% utilization and k is the slope of the curve that varies based on the user application. So, it satisfies $U(0) = 0$ and $U(r_{\text{max}}) = 1$.

Figure 2.1 shows an example of sigmoidal and logarithmic utility functions. It shows three normalized sigmoidal-like utility functions that are expressed by Eq. (2.1) with different parameters $a = 5$, $b = 10$ which is an approximation to a

H. Shajaiah et al., *Resource Allocation with Carrier Aggregation in Cellular Networks*, DOI 10.1007/978-3-319-60540-1_2

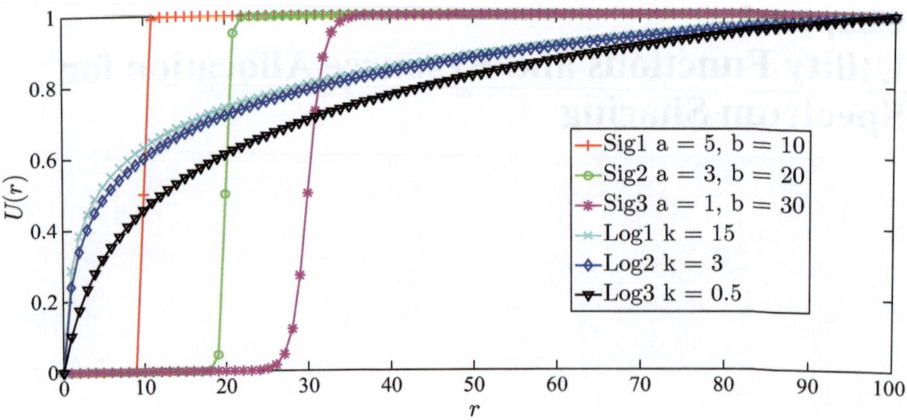

Fig. 2.1 Logarithmic and sigmoidal utility functions $U(r)$ representing delay-tolerant and real-time applications, respectively

step function at rate $r = 10$ (e.g., VoIP), $a = 3, b = 20$ which is an approximation of an adaptive real-time application with inflection point at rate $r = 20$ (e.g., standard definition video streaming), $a = 1, b = 30$ which is also an approximation of an adaptive real-time application with inflection point at rate $r = 30$ (e.g., high definition video streaming). In addition Fig. 2.1 shows three logarithmic functions that are expressed by Eq. (2.2) with $r_{\max} = 100$ and different k parameters which are approximations for delay-tolerant applications (e.g., FTP). We use $k = \{15, 3, 0.5\}$. It is noticeable that real-time applications require a minimum rate, i.e., the inflection point, after that rate the application QoS is fulfilled to a large extent whereas logarithmic utility functions provide some QoS at low rates which is suitable for the delay-tolerant applications nature. Realistic values of a, b, and k for real mobile applications, e.g., youtube and FTP, are shown in [16, 18, 22].

2.2 Utility Proportional Fairness Resource Allocation

In proportional fairness resource allocation model, each user must be allocated some rate. This is guaranteed as allocating zero rate to any user will set the efficiency of the network to zero. The proportional fairness model is presented in the following equation:

$$r_i = \arg\max_{r_i} \prod_{i=1}^{N} U_i(r_i) \tag{2.3}$$

where $U_i(r_i)$ is the utility function of the ith user allocated resource r_i and N is the number of users. The objective function in Eq. (2.3) ensures non-zero resource allocation for all users which guarantees minimum QoS for all users. Frank Kelly algorithm [23] can be used to achieve rate allocation with the fairness model. It achieves Pareto optimal resource allocation across the network while using a proportional fairness approach to distribute all network resources where Pareto optimal or Pareto efficient solutions are the solutions that distribute all of the network resources; i.e., also referred to as the Pareto front [24–26]. Frank Kelly algorithm uses an iterative process to determine the rate that needs to be allocated to each user as well as the price the network should charge each user for the allocated resources. In the next chapters we will be using methods that are based on the Frank Kelly algorithm to solve different proportional fairness resource allocation formulations.

2.3 Utility Proportional Fairness Resource Allocation with Carrier Aggregation

In this book, a utility proportional fairness (UPF) resource allocation optimization framework is proposed to allocate multi-carrier resources optimally among active mobile users from their all in range carriers based on carrier aggregation scenario [27–31]. Throughout the next chapters we present different resource allocation methods for multi-carrier wireless systems. First, we present a multi-stage RA approach which uses a utility proportional fairness RA optimization problem to allocate each carrier resources separately in a multi-stage basis while taking into consideration the resources allocated to each user from other carriers every time the RA optimization problem is executed. The UPF resource allocation optimization problem that we use in the multi-stage RA with CA approach is given by

$$\begin{aligned}
\max_{\mathbf{r}^j} \quad & \prod_{i=1}^{M_j} U_i\left(r_i^j + c_i^j\right) \\
\text{subject to} \quad & \sum_{i=1}^{M_j} r_i^j \le R_j, \\
& r_i^j \ge 0, \quad i = 1, 2, \ldots, M_j, \\
& c_i^j = \sum_{n=1, n\neq j}^{K} v_i^n r_i^{n,\text{opt}}, \\
& v_i^n = \begin{cases} 1, & \text{the } i\text{th UE } \in \mathscr{M}_n \\ 0, & \text{the } i\text{th UE } \notin \mathscr{M}_n \end{cases}
\end{aligned} \tag{2.4}$$

where optimization problem (2.4) is carrier j RA optimization problem, $\mathcal{M}_j$ is the set of users located under the coverage area of the jth eNodeB and $M_j = |\mathcal{M}_j|$ is the number of users in the set $\mathcal{M}_j$, $\mathbf{r}^j = \{r_1^j, r_2^j, \ldots, r_{M_j}^j\}$, R_j is the jth carrier available resources, c_i^j is equivalent to the total rates allocated to the ith user by other carriers in its range, v_i^n is equivalent to 1 if the ith UE $\in \mathcal{M}_n$ and is equivalent to 0 if the ith UE $\notin \mathcal{M}_n$, and $r_i^{n,\text{opt}}$ is the optimal rate allocated to the ith user by the nth carrier.

In order to consider the case when it is required to treat users differently when assigning the network resources, we introduced a user discrimination feature to the resource allocation framework such that certain group of users (e.g., public safety users in systems that consider spectrum sharing between public safety and commercial users) are given priority when allocating the network resources. Furthermore, we developed resource allocation with CA methods to allocate multi-carrier resources based on user discrimination and used UPF optimization problem to calculate the allocated resources.

2.4 MATLAB Code

The following code plots the applications' utility functions shown in Fig. 2.1.

```
n = 0:100;
L=length(n);
%%Sigmoidal Utlity Function with a=5 and b=10%%
U1 = zeros(L,1);
for r = 1:L
    U1(r) = ((1+exp(5*10))/exp(5*10))*((1/(1+exp(-5*(n(r)-10)
        ))-(1/(1+exp(5*10)))));
end
plot(n,U1, '-r+');
hold on
%%Sigmoidal Utility Function with a=3 and b=20%%
U2 = zeros(L,1);
for r = 1:L
    U2(r) = ((1+exp(3*20))/exp(3*20))*((1/(1+exp(-3*(n(r)-20)
        ))-(1/(1+exp(3*20)))));
end
plot(n,U2, '-go');
hold on
%%Sigmoidal Utility Function with a=1 and b=30%%
U3 = zeros(L,1);
for r = 1:L
    U3(r) = ((1+exp(1*30))/exp(1*30))*((1/(1+exp(-1*(n(r)-30)
        ))-(1/(1+exp(1*30)))));
end
plot(n,U3, '-m*');
hold on
%%Logarithmic Utility Function with k=15%%
```

```
U4 = zeros(L,1);
for r = 1:L
    U4(r) = log(1+(15*n(r)))/log(1+(15*100));
end
plot(n,U4, '-cx');
hold on
%%Logarithmic Utility Function with k=3%%
U5 = zeros(L,1);
for r = 1:L
    U5(r) = log(1+(3*n(r)))/log(1+(3*100));
end
plot(n,U5, '-bd');
hold on
%%Logarithmic Utility Function with k=0.5%%
U6 = zeros(L,1);
for r = 1:L
    U6(r) = log(1+(0.5*n(r)))/log(1+(0.5*100));
end
plot(n,U6, '-kv');

ylim([0 1]);
I=legend('Sig1 a = 5, b = 10','Sig2 a = 3, b = 20', 'Sig3 a =
    1, b = 30', 'Log1 k = 15','Log2 k = 3', 'Log3 k = 0.5','
    Interpreter','latex');
set(I,'interpreter','latex');
xlabel('$r$','Interpreter','latex');
ylabel('$U(r)$','Interpreter','latex');
```

References

1. G. Tychogiorgos, A. Gkelias, K.K. Leung, Utility-proportional fairness in wireless networks, in *PIMRC* (IEEE, New York, 2012), pp. 839–844
2. S. Shenker, Fundamental design issues for the future internet. IEEE J. Sel. Areas Commun. **13**, 1176–1188 (1995)
3. Y. Wang, A. Abdelhadi, T.C. Clancy, Optimal power allocation for LTE users with different modulations, in *2016 Annual IEEE Systems Conference (SysCon)*, April 2016, pp. 1–5
4. M. Ghorbanzadeh, A. Abdelhadi, C. Clancy, Distributed resource allocation, in *Cellular Communications Systems in Congested Environments* (Springer, Berlin, 2017), pp. 61–91
5. J.-W. Lee, R.R. Mazumdar, N.B. Shroff, Downlink power allocation for multi-class wireless systems. IEEE/ACM Trans. Netw. **13**, 854–867 (2005)
6. A. Kumar, A. Abdelhadi, T.C. Clancy, A delay efficient multiclass packet scheduler for heterogeneous M2M uplink, in *IEEE MILCOM* (2016)
7. A. Kumar, A. Abdelhadi, T.C. Clancy, An online delay efficient packet scheduler for M2M traffic in industrial automation, in *IEEE Systems Conference* (2016)
8. A. Kumar, A. Abdelhadi, T.C. Clancy, A delay optimal MAC and packet scheduler for heterogeneous M2M uplink. CoRR, abs/1606.06692 (2016)
9. A. Kumar, A. Abdelhadi, T.C. Clancy, A delay-optimal packet scheduler for M2M uplink, in *IEEE MILCOM* (2016)
10. A. Abdelhadi, C. Clancy, A utility proportional fairness approach for resource allocation in 4G-LTE, in *IEEE International Conference on Computing, Networking, and Communications (ICNC), CNC Workshop* (2014)

11. A. Abdelhadi, C. Clancy, A robust optimal rate allocation algorithm and pricing policy for hybrid traffic in 4G-LTE, in *IEEE International Symposium on Personal, Indoor, and Mobile Radio Communications (PIMRC)* (2013)
12. Y. Wang, A. Abdelhadi, A QoS-based power allocation for cellular users with different modulations, in *2016 International Conference on Computing, Networking and Communications (ICNC)*, February 2016, pp. 1–5
13. A. Abdelhadi, H. Shajaiah, Optimal resource allocation for smart phones with multiple applications with MATLAB instructions. Technical Reports (2016)
14. A. Abdelhadi, H. Shajaiah, Application-aware resource allocation with carrier aggregation using MATLAB. Technical Reports (2016)
15. Z. Kbah, A. Abdelhadi, Resource allocation in cellular systems for applications with random parameters, in *2016 International Conference on Computing, Networking and Communications (ICNC)*, February 2016, pp. 1–5
16. A. Abdelhadi, H. Shajaiah, Optimal resource allocation for cellular networks with MATLAB instructions. CoRR, abs/1612.07862 (2016)
17. M. Ghorbanzadeh, A. Abdelhadi, C. Clancy, Utility functions and radio resource allocation, in *Cellular Communications Systems in Congested Environments* (Springer, Berlin, 2017), pp. 21–36
18. M. Ghorbanzadeh, A. Abdelhadi, C. Clancy, Centralized resource allocation, in *Cellular Communications Systems in Congested Environments* (Springer, Berlin, 2017), pp. 37–60
19. M. Ghorbanzadeh, A. Abdelhadi, C. Clancy, Radio resource block allocation, in *Cellular Communications Systems in Congested Environments* (Springer, Berlin, 2017), pp. 117–146
20. M. Ghorbanzadeh, A. Abdelhadi, C. Clancy, Spectrum-shared resource allocation, in *Cellular Communications Systems in Congested Environments* (Springer, Berlin, 2017), pp. 147–178
21. F. Wilson, I. Wakeman, W. Smith, Quality of service parameters for commercial application of video telephony, in *Proceedings of the Human Factors in Telecommunications Symposium* (Darmstadt, 1993), pp. 139–148
22. M. Ghorbanzadeh, A. Abdelhadi, A. Amanna, J. Dwyer, T. Clancy, Implementing an optimal rate allocation tuned to the user quality of experience, in *Computing, Networking and Communications (ICNC), 2015 International Conference on*, February 2015, pp. 292–297
23. F. Kelly, A. Maulloo, D. Tan, Rate control in communication networks: shadow prices, proportional fairness and stability. J. Oper. Res. Soc. **49**, 237–252 (1998)
24. R. Madan, S. Boyd, S. Lall, Fast algorithms for resource allocation in wireless cellular networks. IEEE/ACM Trans. Netw. **18**, 973–984 (2010)
25. J.B. Taylor, *Principles of Microeconomics*. Microeconomics Series (Houghton Mifflin, Boston, 1998)
26. S. Boyd, L. Vandenberghe, *Introduction to Convex Optimization with Engineering Applications*. Convex Optimization, Course Reader for EE364 (1999)
27. M. Iwamura, K. Etemad, M.-H. Fong, R. Nory, R. Love, Carrier aggregation framework in 3GPP LTE-advanced [WiMAX/LTE Update]. IEEE Commun. Mag. **48**(8), 60–67 (2010)
28. Y. Wang, K.I. Pedersen, T.B. Sørensen, P.E. Mogensen, Utility maximization in LTE-advanced systems with carrier aggregation, in *VTC Spring* (2011), pp. 1–5
29. G. Yuan, X. Zhang, W. Wang, Y. Yang, Carrier aggregation for LTE-advanced mobile communication systems. IEEE Commun. Mag. **48**, 88–93 (2010)
30. Z. Shen, A. Papasakellariou, J. Montojo, D. Gerstenberger, F. Xu, Overview of 3GPP LTE-advanced carrier aggregation for 4G wireless communications. IEEE Commun. Mag. **50**, 122–130 (2012)
31. *Frequency Spectrum Wall Chart* (Commerce Dept., National Telecommunications and Information Administration, Office of Spectrum Management, 2016)

Chapter 3
Multi-Stage Resource Allocation with Carrier Aggregation

In this chapter, we present a resource allocation with carrier aggregation optimization problem to allocate the eNodeB's carrier resources optimally among users in its coverage area while taking into consideration the rates allocated to each user from other carriers for cellular system [3–5]. We propose two multi-stage resource allocation with carrier aggregation approaches. The first approach uses distributed (decentralized) multi-stage algorithms to allocate users, under the coverage area of a primary carrier and a secondary carrier, the resources from both carriers. The second approach uses centralized multi-stage algorithms to allocate users the resources optimally from all in band carriers and give each user the ability to select its primary and secondary carriers based on their offered prices in order to provide a minimum price for the allocated resources.

3.1 Distributed Resource Allocation with Carrier Aggregation

In this section, we present a distributed multi-stage approach for allocating primary and secondary carriers resources optimally among mobile users based on carrier aggregation. We focus on finding an optimal solution for the carrier aggregation resource allocation problem for a group of users running two types of applications presented by logarithmic utility functions or sigmoidal-like utility functions. These utility functions are concave and non-concave utility functions, respectively. The RA optimization problem assigns part of the bandwidth from two carriers to each user subscribing for a mobile service taking into consideration that each user is

---

The content in this chapter is reproduced with permission after modifications (License numbers 4078230506280 and 4078230239335). For the original article, please refer to [1, 2].

H. Shajaiah et al., *Resource Allocation with Carrier Aggregation in Cellular Networks*, DOI 10.1007/978-3-319-60540-1_3

getting a minimum QoS. Our objective is to allocate the resources from two carriers to each user based on its application that is represented by a utility function. The contributions in this section are summarized as:

- We present a resource allocation optimization problem with carrier aggregation that gives priority to real-time application users when allocating resources.
- We prove that the optimal rate allocated by the two carriers to each user when using carrier aggregation is equivalent to the optimal rate allocated to the same user by one carrier that has resources equivalent to the total resources in the two carriers. We present a two-stage carrier aggregation rate allocation algorithm to solve the optimization problem and present its corresponding simulation results.

3.1.1 Problem Formulation

We consider two eNodeBs that have the same coverage area and M UEs. One of the eNodeBs is considered to be the primary carrier and the other one is the secondary carrier. Each user is allocated certain bandwidth r_i based on the type of application the UE is running. Our goal is to determine the optimal bandwidth that needs to be allocated to each user by the two eNodeBs.

We assume the utility functions $U_i(r_i)$ to be strictly concave or sigmoidal-like functions [6–14, 16]. Logarithmic utility functions expressed by Eq. (2.2) and sigmoidal-like utility functions expressed by Eq. (2.1) are used to represent delay-tolerant and real-time applications, respectively, as in [7, 15–22].

3.1.1.1 Single-Carrier Optimization Problem

The basic formulation of a single-carrier resource allocation problem is given by the following optimization problem:

$$\begin{aligned} \max_{\mathbf{r}_{\text{single}}} \quad & \prod_{i=1}^{M} U_i(r_{i,\text{single}}) \\ \text{subject to} \quad & \sum_{i=1}^{M} r_{i,\text{single}} \leq R, \\ & r_{i,\text{single}} \geq 0, \quad i = 1, 2, \ldots, M \end{aligned} \tag{3.1}$$

where R is the maximum achievable rate of the eNodeB, $r_{i,\text{single}}$ is the rate for user i, and M is the number of UEs.

The optimization problem (3.1) is a convex optimization problem [23–25] and there exists a unique tractable global optimal solution [8]. The objective function in the optimization problem (3.1) is equivalent to $\max_{\mathbf{r}_{\text{single}}} \sum_{i=1}^{M} \log U_i(r_{i,\text{single}})$. The solution of this optimization problem is the global optimal solution for the resource allocation problem when resources are allocated by one eNodeB.

For the carrier aggregation resource allocation case, the optimization problem is divided into two stages as shown in Sect. 3.1.2.

3.1.2 *Primary and Secondary Carriers Optimization Problem*

3.1.2.1 Primary Carrier

The two carriers optimization problem is done in two stages, primary and secondary stages.

The optimization problem for the first carrier can be written as:

$$
\begin{aligned}
\max_{\mathbf{r_p}} \quad & \prod_{i=1}^{M} U_i(r_{i,p}) \\
\text{subject to} \quad & \sum_{i=1}^{M} r_{i,p} \leq R_p, \\
& r_{i,p} \geq 0, \quad i = 1, 2, \ldots, M
\end{aligned}
\tag{3.2}
$$

where $\mathbf{r_p} = \{r_{1,p}, r_{2,p}, \ldots, r_{M,p}\}$, M is the number of UEs in the coverage area of primary user eNodeB, and R_p is the maximum achievable rate of the primary carrier. The resource allocation objective function is to maximize the total system utility when allocating resources to each user. Furthermore, it provides proportional fairness among utilities. Users running real-time applications are allocated more resources in this approach.

The optimization problem (3.2) is a convex optimization problem and there exists a unique tractable global optimal solution [8]. The objective function in the optimization problem (3.2) is equivalent to $\max_{\mathbf{r_p}} \sum_{i=1}^{M} \log U_i(r_{i,p})$. The solution of this optimization problem is the first optimal solution that gives each of the M users the optimal rate $r_{i,p}^{\text{opt}}$ only from the primary carrier and not yet the final optimal rate.

3.1.2.2 Secondary Carrier

As mentioned before, we consider a secondary carrier eNodeB located in the same coverage area of the same mobile system. Again, M is the number of mobile users in the coverage area. Once the primary carrier finishes allocating its resources to the M users, the secondary carrier starts to allocate its resources to the same users while ensuring a minimum user QoS. Therefore, we assume again that the secondary carrier will allocate the resources based on utility proportional fairness.

The optimization problem for the secondary carrier can be written as:

$$
\begin{aligned}
&\max_{\mathbf{r_s}} && \prod_{i=1}^{M} U_i\left(r_{i,s} + r_{i,p}^{\text{opt}}\right) \\
&\text{subject to} && \sum_{i=1}^{M} r_{i,s} \leq R_s, \\
& && r_i \geq 0, \quad i = 1, 2, \ldots, M
\end{aligned}
\tag{3.3}
$$

where $\mathbf{r_s} = \{r_{1,s}, r_{2,s}, \ldots, r_{M,s}\}$ is the rate for user i, R_s is the maximum achievable rate by the secondary carrier, and $r_{i,p}^{\text{opt}}$ is the first optimal rate allocated to user i by the primary carrier and estimated in (3.2). The optimization problem here gives priority to real-time application users and ensures that the minimum allocated rate for each user equals to the first optimal rate $r_{i,p}^{\text{opt}}$ estimated in (3.2).

The optimization problem (3.3) is a convex optimization problem and there exists a unique tractable global optimal solution [8]. The objective function in the optimization problem (3.3) is equivalent to $\max_{\mathbf{r_s}} \sum_{i=1}^{M} \log U_i(r_{i,s} + r_{i,p}^{\text{opt}})$. The global optimal rate for each user is obtained by the sum of the solution given by (3.2) $r_{i,p}^{\text{opt}}$ and the solution given by (3.3) $r_{i,s}^{\text{opt}}$ for user i and is equal to $r_{i,\text{agg}}^{\text{opt}} = r_{i,s}^{\text{opt}} + r_{i,p}^{\text{opt}}$, such that $r_{i,\text{agg}}^{\text{opt}}$ is the global optimal solution that gives each of the M users the optimal rate from both the primary and secondary carriers and considered the final optimal rate.

3.1.2.3 Equivalence

In this section, we show the equivalence of the optimal rate $r_{i,\text{agg}}^{\text{opt}}$ given to each user by the primary and secondary eNodeBs to the optimal rate given to the same user by a single eNodeB, given by the single-carrier optimization problem (3.1), when its available resources are equivalent to the resources available in both the primary and secondary eNodeBs in the carrier aggregation case.

Theorem 3.1.1 *The optimal rate $r_{i,\text{agg}}^{\text{opt}}$ allocated to user i by the two carriers from optimization problem (3.2) and optimization problem (3.3) is equivalent to the optimal rate allocated to the same user by the single-carrier optimization problem (3.1) when $R = R_p + R_s$.*

Proof From the optimization problem (3.2), we have the Lagrangian:

$$
L_p(r_{i,p}) = \left(\sum_{i=1}^{M} \log U_i(r_{i,p})\right) - P_p\left(\sum_{i=1}^{M} r_{i,p} - R_p - z_p\right), \tag{3.4}
$$

where $z_p \geq 0$ is the slack variable and P_p is the Lagrange multiplier which is equivalent to the shadow price that corresponds to the total price per unit bandwidth for the M channels as in [8]. So, we have

$$\frac{\partial L_p(r_{i,p})}{\partial r_{i,p}} = \frac{U_i'(r_{i,p})}{U_i(r_{i,p})} - P_p = 0, \tag{3.5}$$

solving for $r_{i,p}$ we obtain $r_{i,p}^{\text{opt}}$.

From optimization problem (3.3), we have the Lagrangian:

$$L_s(r_{i,s}) = \left(\sum_{i=1}^{M} \log U_i\left(r_{i,s} + r_{i,p}^{\text{opt}}\right)\right) - P_s\left(\sum_{i=1}^{M} r_{i,s} - R_s - z_s\right), \tag{3.6}$$

where $z_s \geq 0$ is the slack variable and P_s is the Lagrange multiplier. So, we have

$$\frac{\partial L_s(r_{i,s})}{\partial r_{i,s}} = \frac{U_i'\left(r_{i,s} + r_{i,p}^{\text{opt}}\right)}{U_i\left(r_{i,s} + r_{i,p}^{\text{opt}}\right)} - P_s = 0, \tag{3.7}$$

solving for $r_{i,s}$ we obtain $r_{i,s}^{\text{opt}}$. Replacing $r_{i,s} + r_{i,p}^{\text{opt}}$ in Eq. (3.6) by a new variable $r_{i,\text{agg}}$ such that $r_{i,\text{agg}} = r_{i,s} + r_{i,p}^{\text{opt}}$ and rewriting the Lagrangian in terms of $r_{i,\text{agg}}$, we obtain

$$L_{\text{agg}}(r_{i,\text{agg}}) = \left(\sum_{i=1}^{M} \log U_i(r_{i,\text{agg}})\right) - P_s\left(\sum_{i=1}^{M} \left(r_{i,\text{agg}} - r_{i,p}^{\text{opt}}\right) - R_s - z_s\right), \tag{3.8}$$

where $r_{i,\text{agg}} \geq r_{i,p}^{\text{opt}}$. From the primary carrier, we have $\sum_{i=1}^{M} r_{i,p}^{\text{opt}} = R_p$. So, Eq. (3.8) is equivalent to

$$L(r_{i,\text{agg}}) = \sum_{i=1}^{M} \log U_i(r_{i,\text{agg}}) - P_s\left(\sum_{i=1}^{M} r_{i,\text{agg}} - R - z_s\right). \tag{3.9}$$

From problem (3.1), we have

$$L_{\text{single}}(r_{i,\text{single}}) = \left(\sum_{i=1}^{M} \log U_i(r_{i,\text{single}})\right) - P\left(\sum_{i=1}^{M} (r_{i,\text{single}} - R - z)\right), \tag{3.10}$$

equivalent to (3.8) for $r_i \geq r_{i,p}^{\text{opt}}$. Therefore, the optimal solution $r_{i,\text{agg}}^{\text{opt}}$ given by (3.8) is equivalent to the optimal solution $r_{i,\text{single}}^{\text{opt}}$ given by (3.10) when $R = R_p + R_s$.

Algorithm 1 First stage of UE carrier aggregation algorithm

Send initial bid $w_{i,p}(1)$ to eNodeB
loop
  Receive shadow price $P_p(n)$ from eNodeB
  if STOP from eNodeB **then**
    Calculate allocated rate $r_{i,p}^{\text{opt}} = \frac{w_{i,p}(n)}{P_p(n)}$
  else
    Solve $r_{i,p}(n) = \arg\max_{r_{i,p}} \Big(\log U_i(r_{i,p}) - P_p(n) r_{i,p} \Big)$
    Send new bid $w_{i,p}(n) = P_p(n) r_{i,p}(n)$ to eNodeB
  end if
end loop

Algorithm 2 First-stage eNodeB carrier aggregation algorithm

---

loop
  Receive bids $w_{i,p}(n)$ from UEs {Let $w_{i,p}(0) = 0 \ \forall i$}
  if $|w_{i,p}(n) - w_{i,p}(n-1)| < \delta \ \forall i$ **then**
    STOP and allocate rates (i.e., $r_{i,p}^{\text{opt}}$ to user i)
  else
    Calculate $P_p(n) = \frac{\sum_{i=1}^{M} w_{i,p}(n)}{R_p}$
    Send new shadow price $P_p(n)$ to all UEs
  end if
end loop

Algorithm 3 Second-stage UE carrier aggregation algorithm

Send initial bid $w_{i,s}(1)$ to eNodeB
loop
  Receive shadow price $P_s(n)$ from eNodeB
  if STOP from eNodeB **then**
    Calculate allocated rate $r_{i,s}^{\text{opt}} = \frac{w_{i,s}(n)}{P_s(n)}$
  else
    Solve $r_{i,s}(n) = \arg\max_{r_{i,s}} \Big(\log U_i(r_{i,s} + r_{i,p}^{\text{opt}}) - P_s(n) r_i \Big)$
    Send new bid $w_{i,s}(n) = P_s(n) r_{i,s}(n)$ to eNodeB
  end if
end loop

3.1.3 *Multi-Stage Distributed Resource Allocation Algorithm*

The distributed resource allocation algorithm is based on utility proportional fairness policy. The algorithm is performed in multiple stages based on the number of carriers. For the case of one primary carrier and one secondary carrier, the algorithm is divided into two stages. Algorithms 1 and 2 are the UE and the eNodeB algorithms, respectively, performed at the first stage. Algorithms 3 and 4 are the UE and the eNodeB algorithms, respectively, performed at the second stage. In the first stage, each UE transmits an initial bid $w_{i,p}(1)$ to the primary eNodeB. The eNodeB checks whether the difference between the current received bid and the previous one

Algorithm 4 Second-stage eNodeB carrier aggregation algorithm

loop
  Receive bids $w_{i,s}(n)$ from UEs {Let $w_{i,s}(0) = 0 \ \forall i$}
  if $|w_{i,s}(n) - w_{i,s}(n-1)| < \delta \ \forall i$ **then**
    STOP and allocate rates (i.e., $r_{i,s}^{\text{opt}}$ to user i)
  else
    Calculate $P_s(n) = \frac{\sum_{i=1}^{M} w_{i,s}(n)}{R_s}$
    Send new shadow price $P_s(n)$ to all UEs
  end if
end loop

is less than a threshold δ, if so it exits. Otherwise, if the difference is greater than δ, eNodeB calculates the shadow price $P_p(n) = \frac{\sum_{i=1}^{M} w_{i,p}(n)}{R_p}$. The shadow price does not depend on the number of users competing for the resources, it only depends on users' bids and the eNodeB's available resources. The estimated $P_p(n)$ is then sent to the UE where it is used to calculate the rate $r_{i,p}(n)$ which is the solution of the optimization problem $r_{i,p}(n) = \arg\max_{r_{i,p}}(\log U_i(r_{i,p}) - P_p(n) r_{i,p})$. A new bid $w_{i,p}(n)$ is calculated using $r_{i,p}(n)$ where $w_{i,p}(n) = P_p(n) r_{i,p}(n)$. All UEs send their new bids $w_{i,p}(n)$ to the primary eNodeB. The first stage of the algorithm is finalized by the primary eNodeB. Each UE then calculates its allocated rate $r_{i,p}^{\text{opt}} = \frac{w_{i,p}(n)}{P_p(n)}$.

After allocating rates from primary carrier, the second stage of the algorithm starts performing. Each UE transmits an initial bid $w_{i,s}(1)$ to the secondary eNodeB. The secondary eNodeB checks whether the difference between the current received bid and the previous one is less than a threshold δ, if so it exits. Otherwise, if the difference is greater than δ, the secondary eNodeB calculates the shadow price $P_s(n) = \frac{\sum_{i=1}^{M} w_{i,s}(n)}{R_s}$. The estimated price $P_s(n)$ is then sent to the UE where it is used to calculate the rate $r_{i,s}(n)$ which is the solution of the optimization problem $r_{i,s}(n) = \arg\max_{r_{i,s}}(\log U_i(r_{i,s} + r_{i,p}^{\text{opt}}) - P_s(n) r_{i,s})$. A new bid $w_{i,s}(n)$ is calculated using $r_{i,s}(n)$ where $w_{i,s}(n) = P_s(n) r_{i,s}(n)$. All UEs send their new bids $w_{i,s}(n)$ to the secondary eNodeB. The second stage of the Algorithm is finalized by the secondary eNodeB. Each UE then calculates its allocated rate $r_{i,s}^{\text{opt}} = \frac{w_{i,s}(n)}{P_s(n)}$. A flow diagram of the aforementioned multi-stage distributed RA with CA algorithm is shown in Fig. 3.1.

3.1.4 Optimal Rate Allocation Numerical Results

As shown in Fig. 3.2, we consider two eNodeBs with the same coverage area and six UEs. One of the eNodeBs is the primary carrier and the other one is the secondary carrier with a coverage area that is almost the same for the two carriers.

Fig. 3.1 Flow diagram of multi-stage distributed algorithm

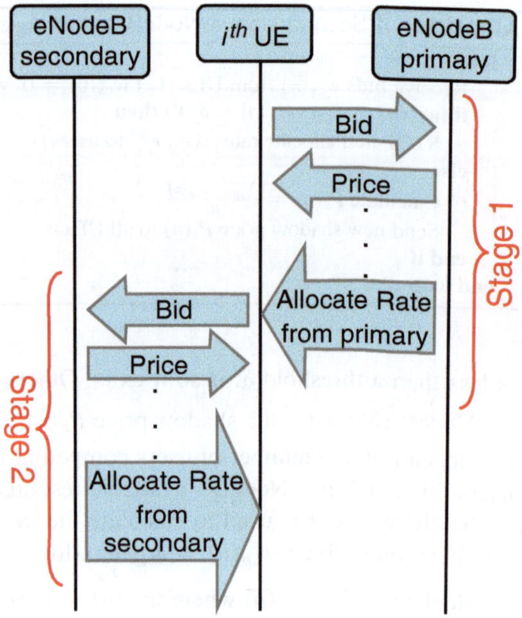

Fig. 3.2 System model for primary and secondary carriers with almost the same coverage area

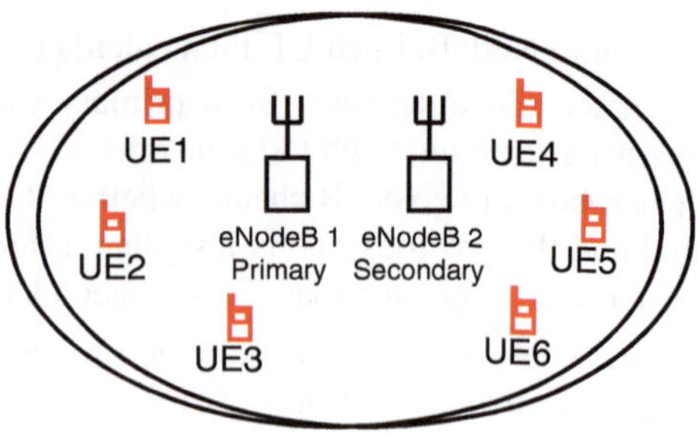

The algorithms presented in Sect. 3.1.3 were applied to various sigmoidal-like and logarithmic utility functions in C++ (Sect. 3.4.1). Simulation results showed convergence to the optimal global point. In Fig. 3.3, we show three normalized sigmoidal-like utility functions expressed in Eq. (2.1), each one is corresponding to a user application running on one UE. We use different parameters a and b for each one where $a = 5$, $b = 10$ for the first user, $a = 3$, $b = 20$ for the second user, and $a = 1$, $b = 30$ for the third user. Each sigmoidal-like function is an approximation to a step function at rate b. We also show three logarithmic functions expressed in Eq. (2.2), which represent delay-tolerant applications, with $k = \{15, 3, 0.5\}$ for users 4, 5, and 6, respectively. We set $r_{\max} = 120$.

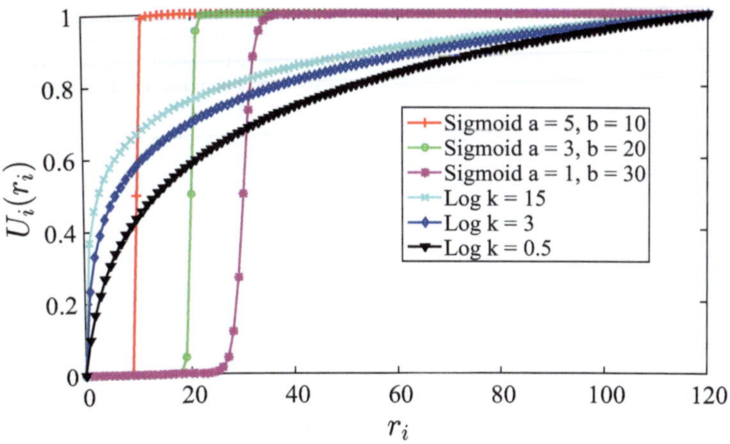

Fig. 3.3 The users utility functions $U_i(r_i)$

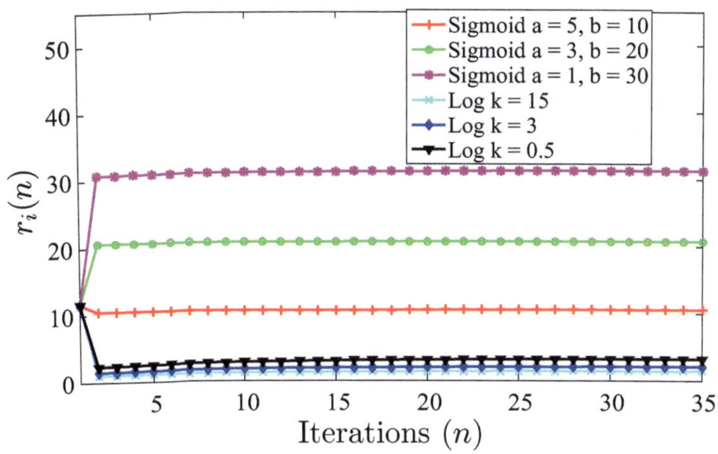

Fig. 3.4 The rates $r_{i,p}(n)$ with the number of iterations n for different users and $R_p = 70$

3.1.4.1 Convergence Dynamics for the First Stage of the Algorithm

We applied Algorithms 1 and 2 (i.e., the first-stage algorithms) in C++ to the sigmoidal-like and logarithmic utility functions shown in Fig. 3.3. We set $R_p = 70$ and $\delta = 10^{-2}$. In Fig. 3.4, we show the simulation results for the rate of different users and the number of iterations. As mentioned before, the sigmoidal-like utility functions are given priority over the logarithmic utility functions for rate allocation and this explains the results we got in Fig. 3.4 where the steady-state rate of each sigmoidal-like function exceeds the inflection point b_i. In Fig. 3.5, we show the bids of the six users with the number of iterations. As expected, the higher the user bids, the higher the allocated rate is for that user. The algorithm allows users with real-

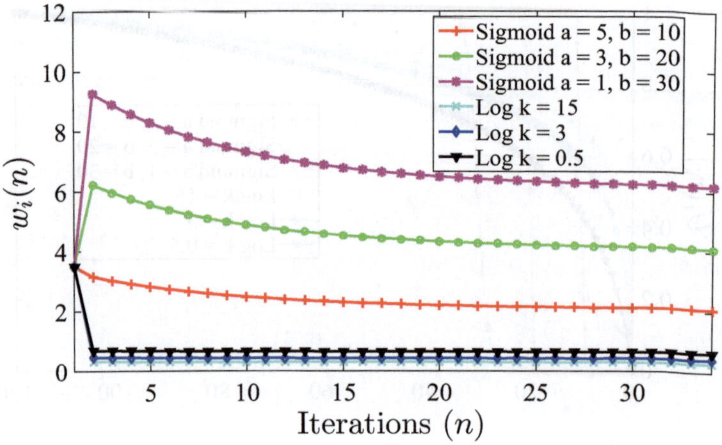

Fig. 3.5 The bids convergence $w_{i,p}(n)$ with the number of iterations n for different users and $R_p = 70$

time applications, presented in sigmoidal-like utility functions, to bid higher than the other users until each one of them reaches its inflection point then the elastic traffic starts dividing the remaining resources among them based on their parameters. The first optimal rates for the six users $r_{i,p}^{\text{opt}} = \{10.64, 20.88, 31.41, 1.54, 2.19, 3.26\}$ are obtained at the end when Algorithms 1 and 2 are executed. The first optimal rates are used in the next simulation that is performed for the secondary eNodeB and the same six UEs.

3.1.4.2 Convergence Dynamics for the Second Stage of the Algorithm

We applied Algorithms 3 and 4 in C++ to the sigmoidal-like and logarithmic utility functions. We set $R_s = 50$ and $\delta = 10^{-2}$.

In Fig. 3.6, we show the simulation results for the rate of the six users and the number of iterations. Again, the sigmoidal-like utility functions are given priority over the logarithmic utility functions for rate allocation, but since each sigmoidal-like function reached its steady state in the first stage of the Algorithm most of the secondary carrier resources (i.e., R_s) is distributed among the logarithmic functions. In the second stage, the optimal rates for the real-time application users $r_{i,s}^{\text{opt}}$ slightly increased from the first optimal rate $r_{i,p}^{\text{opt}}$ as they were given priority to reach their optimal rates in the first stage by the primary eNodeB, whereas users' applications with elastic traffic divided the remaining resources among them and showed a high increase in their second optimal rate $r_{i,s}^{\text{opt}}$ from their first optimal rates obtained in the first stage. The optimal rates obtained at the end of the second stage are $r_{i,s}^{\text{opt}} = \{0.51, 0.88, 2.735, 10.94, 14.06, 21.87\}$.

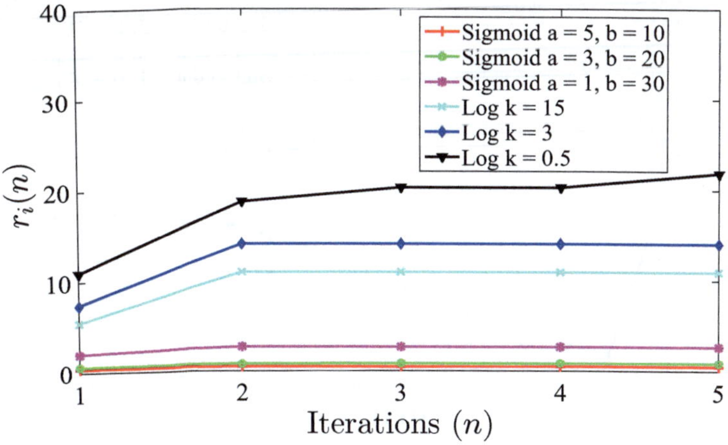

Fig. 3.6 The rates $r_{i,s}(n)$ with the number of iterations n for different users and $R_s = 50$

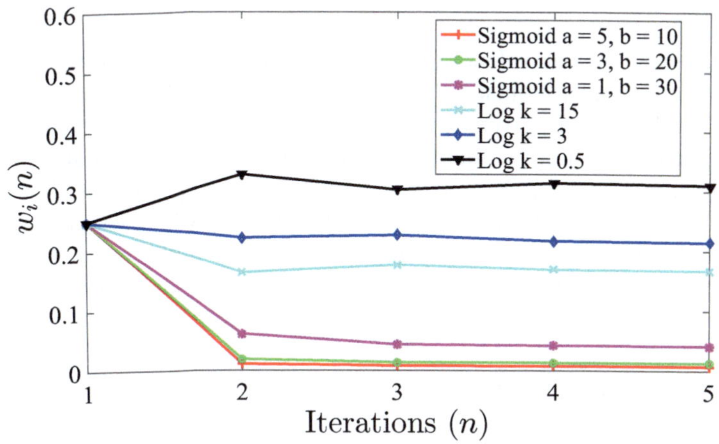

Fig. 3.7 The bids convergence $w_{i,s}(n)$ with the number of iterations n for different users and $R_s = 50$

In Fig. 3.7, we show the bids of the six users with the number of iterations. As expected, the higher the user bids, the higher the allocated rate is for that user. The algorithm allows users with real-time applications, presented in sigmoidal-like utility functions, to bid higher than the other users until each one of them reaches its inflection point, but since these users reached their steady states in stage 1 of the algorithm the elastic traffic users bid higher than the inelastic traffic users and share the secondary carrier's resources among them based on their parameters.

The final optimal rate for each user $r_{i,agg}^{opt}$ is the sum of $r_{i,p}^{opt}$ obtained at the end of the first stage of the algorithm and $r_{i,s}^{opt}$ obtained at the end of the second stage. As expected, the final optimal rates for the six users sum up to 120 which is the total rate of the primary and secondary maximum rates.

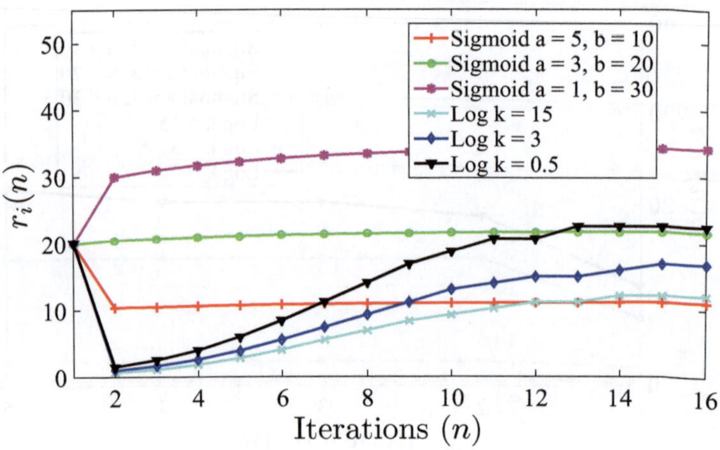

Fig. 3.8 The rates $r_{i,\text{single}}(n)$ with the number of iterations n for different users for the single-carrier case with $R = 120$

We then show the equivalence of the optimal rate $r_{i,\text{agg}}^{\text{opt}}$ allocated to each user by both the primary and secondary eNodeBs to the optimal rate allocated to the same user by a single eNodeB, given by the single carrier when its available resources are equivalent to the resources available in both the primary and secondary eNodeBs. In other words, we show the equivalence of optimal rate $r_{i,\text{single}}^{\text{opt}}$ with $r_{i,p}^{\text{opt}} + r_{i,s}^{\text{opt}}$ when $R = R_p + R_s$. Figure 3.8 shows the optimal rates obtained when we run the first-stage algorithms for the same six users sharing resources of a single carrier with $R = 120$. We made $R_p = R$, $r_{i,p}(n) = r_{i,\text{single}}(n)$, $w_{i,p}(n) = w_i(n)$, and $P_p(n) = P(n)$ when running Algorithms 1 and 2 for the single-carrier case. The optimal rates obtained in this case are $r_{i,\text{single}}^{\text{opt}} = \{11.16, 21.74, 34.22, 13.12, 16.87, 22.50\}$, and they are almost similar to the final optimal rates $r_{i,\text{agg}}^{\text{opt}}$ in the carrier aggregation case when the same users share the resources of two carriers one being the primary and the other being the secondary with a total R_p and R_s of 120.

3.1.4.3 Impact of Dynamic User Activities in the Convergence of the RA Algorithms

We investigate the sensitivity of the proposed resource allocation algorithms to users arrival, departures, and user utility changes due to application activity changes. Let $\mathcal{M}$ represent the set of original users that the eNodeB has started calculating their optimal rates by running the RA algorithms before users arrival or departure occur, and let w_i^{opt} represent the optimal bidding values that correspond to the optimal rates calculated by the RA algorithms. Let $\mathcal{M}'$ represent the new set of users after users departure or arrival. We compare the number of iterations it takes for the algorithms to converge when changing users activities for the two cases described below:

- Case 1: The eNodeB uses the optimal bidding values w_i^{opt}, determined after the convergence of the RA algorithms for the original users, as initial bidding values (i.e., $w_i' = w_i^{\text{opt}}$) for common bidders in $\mathcal{M}'$ and $\mathcal{M}$ when it starts running the RA algorithms to determine the optimal rates for users in $\mathcal{M}'$ after the changes in users' activities.
- Case 2: Cold start, the eNodeB and all active UEs start running the RA algorithms without using or taking into consideration the optimal bids determined by the algorithms, before the changes in users activities, for common users in $\mathcal{M}'$ and $\mathcal{M}$.

We considered the same six UEs ($|\mathcal{M}| = 6$) with the same simulation setup described above. We ran the resource allocation with CA algorithms for the six users and observed the number of iterations that takes for the algorithms to converge to the optimal rates allocated from the eNodeB's primary resources as well as the eNodeB's secondary resources. On the other hand, we considered the arrival of additional two users where the number of users subscribing for mobile services changed from six users to eight users ($|\mathcal{M}'| = 8$) with user 7 running real-time application represented by sigmoidal utility function with $a = 5, b = 10$, and user 8 is running delay-tolerant application represented by logarithmic utility function with parameter $k = 15$. For the two cases described above, we compared the number of iterations that took for the algorithms to converge when the number of users changed to 8. The observed number of iterations in case 1 (when considering the optimal bidding values of the first six users) and case 2 (cold start) are almost the same. In addition, we consider users departure where after the convergence of the algorithms for the original six users, two users (user 5 and user 6) departed and are no longer active for the eNodeB. Again, for the two cases described above, we compared the number of iterations it took for the algorithms to converge. We noticed that the number of iterations is almost the same for the two cases.

For common users in $\mathcal{M}$ and $\mathcal{M}'$, when using the users bidding values w_i^{opt}, that correspond to the optimal rates r_i^{opt} calculated by the eNodeB and UE algorithms when considering all users in $\mathcal{M}$, as initial bidding values w_i' for determining new optimal allocated rates by the RA algorithms, the number of iterations required for the convergence of the algorithms during the process of calculating the optimal rates for the updated users set $\mathcal{M}'$ is not necessarily less than the number of iterations required for the convergence of the algorithms when common UEs in $\mathcal{M}$ and $\mathcal{M}'$ send new bidding values $w_i' \neq w_i^{\text{opt}}$. This is because the optimal rates calculated by the algorithms before users departure or arrival are no longer optimal and new optimal rates will be calculated by the algorithms for users in $\mathcal{M}'$. Therefore, in situations of dynamic users activities, it does not matter whether the algorithms use the latest calculated bidding values or new ones (cold start) for users who did not change their running applications and are still active; that is, the algorithms will not necessarily converge faster when using the latest calculated bidding values before the changes in users activities.

3.2 Centralized Resource Allocation with Carrier Aggregation

In this section, we present a centralized multi-stage approach for allocating multiple carriers' resources among mobile users based on carrier aggregation. We formulate the RA with CA problem into a convex optimization framework. We use logarithmic utility functions to represent delay-tolerant applications and sigmoidal-like utility functions to represent real-time applications running on the UEs subscribing for a mobile service. The primary and secondary carriers optimization problems assign part of the bandwidth from the multiple carriers to each user. A minimum QoS is guaranteed for each user by using a proportional fairness approach. Our objective is to allocate multiple carriers resources optimally among users in their coverage area while giving the user the ability to select the carrier with the lowest price to be its primary carrier and the others to be its secondary carriers. This mechanism allows users to improve their allocated rates by using the CA feature while maintaining the lowest possible price for their allocated aggregated rates. Additionally, our centralized algorithm is performed mostly in the eNodeBs which reduces the transmission overhead created by the distributed algorithm introduced in [1]. The contributions in this section are summarized as:

- We present a resource allocation optimization problem with carrier aggregation that solves for logarithmic and sigmoidal-like utility functions.
- We propose a price selective centralized RA with CA algorithm to allocate multiple carriers resources optimally among users.
- We show that our algorithm is a robust one that converges to the optimal rates whether the eNodeBs available resources are abundant or scarce. We present simulation results for the performance of our resource allocation algorithm.

3.2.1 *Multi-Carrier System Model*

We consider an LTE mobile system with M users and K carriers eNodeBs, one eNodeB in each cell, as illustrated in Fig. 3.9. The users located under the coverage area of the ith eNodeB are forming a set of users $\mathcal{M}_i$ where $\mathcal{M}_i \in \{\mathcal{M}_1, \mathcal{M}_2, \ldots, \mathcal{M}_K\}$ and $M_i = |\mathcal{M}_i|$ is the number of users in the users set $\mathcal{M}_i$ under the coverage area of the ith eNodeB. Each joint user j is located under the coverage area of a set of eNodeBs, as shown in Fig. 3.9, that is given by $\mathcal{K}_j$ where $\mathcal{K}_j \in \{\mathcal{K}_1, \mathcal{K}_2, \ldots, \mathcal{K}_M\}$ and $K_j = |\mathcal{K}_j|$ is the number of eNodeBs in the set $\mathcal{K}_j$ of all in range eNodeBs for user j.

Each eNodeB calculates its offered price per unit bandwidth (assuming that it is the primary carrier for all users under its coverage area) and provides each user under its coverage area with its offered price. Each joint user selects the carrier with the least offered price to be its primary carrier and the rest of all in range carriers to be its secondary carriers. The eNodeB with the least offered price first allocates

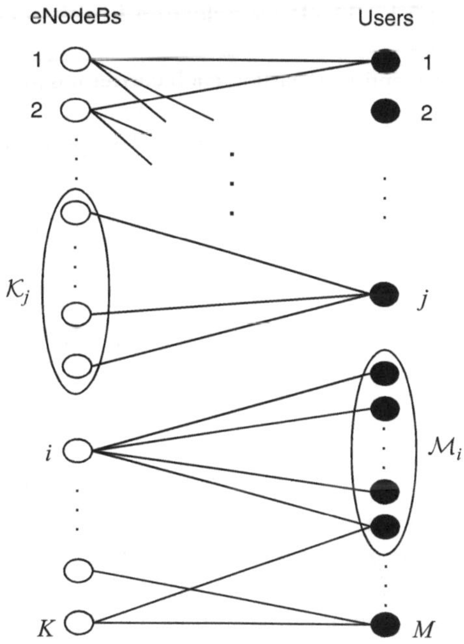

Fig. 3.9 System model for an LTE mobile system with M users and K carriers eNodeBs. $\mathcal{M}_i$ represents the set of users located under the coverage area of the ith eNodeB and $\mathcal{K}_j$ represents the set of all in range eNodeBs for the jth user

its resources to all users under its coverage area based on the applications running on their UEs. The remaining eNodeBs then start allocating their resources in the order of their offered prices to all users under their coverage area based on the users applications and the rates that are allocated to the joint users from other eNodeBs (with lower offered prices).

We express the user satisfaction with its provided service using utility functions [7, 15, 16]. We assume that the jth user's application utility function $U_j(r_j)$ is strictly concave or sigmoidal-like function where r_j is the rate allocated to the jth user. Delay-tolerant applications are represented by logarithmic utility functions expressed by Eq. (2.2) whereas real-time applications are represented by sigmoidal-like utility functions expressed by Eq. (2.1).

3.2.2 *Multi-Carrier Optimization Problem*

In this section, we formulate the RA problem for allocating the primary and secondary carriers resources optimally among users under their coverage areas. Each carrier first calculates its offered price per unit bandwidth assuming that it is the primary carrier for all UEs under its coverage area. Then, each carrier starts allocating its available resources optimally among all users in its coverage area in the order of the carrier's offered price, such that the carrier with a lower offered price performs the RA prior to the one with a higher offered price.

3.2.2.1 The Price Selection Problem and eNodeB Sorting

As mentioned earlier, each carrier calculates its offered price assuming that it is the primary carrier for all users under its coverage area. The carrier's offered price is obtained from the following RA optimization problem:

$$\begin{aligned} \max_{\mathbf{r_i}} \quad & \prod_{j=1}^{M_i} U_j(r_{i,j}) \\ \text{subject to} \quad & \sum_{j=1}^{M_i} r_{i,j} \leq R_i, \\ & r_{i,j} \geq 0, \quad j = 1, 2, \ldots, M_i \end{aligned} \tag{3.11}$$

where $\mathbf{r_i} = \{r_{i,1}, r_{i,2}, \ldots, r_{i,M_i}\}$, M_i is the number of UEs under the coverage area of the ith eNodeB, and R_i is the maximum achievable rate of the ith eNodeB. The resource allocation objective function is to maximize the total system utility when allocating the eNodeB resources. Furthermore, it provides proportional fairness among utilities. Therefore, no user is allocated zero resources and a minimum QoS is provided to each user. Real-time applications are given priority when allocating the eNodeB resources using this approach. Optimization problem (3.11) is a convex optimization problem and there exists a unique tractable global optimal solution [8]. The objective function in optimization problem (3.11) is equivalent to $\max_{\mathbf{r_i}} \sum_{j=1}^{M_i} \log U_j(r_{i,j})$.

From optimization problem (3.11), we have the Lagrangian:

$$L_i(r_{i,j}) = \left(\sum_{j=1}^{M_i} \log U_j(r_{i,j}) \right) - p_i^{\text{offered}} \left(\sum_{j=1}^{M_i} r_{i,j} - R_i - z_i \right), \tag{3.12}$$

where $z_i \geq 0$ is the slack variable and p_i^{offered} is the Lagrange multiplier which is equivalent to the shadow price that corresponds to the ith carrier price per unit bandwidth for the M_i channels as in [8]. The set of all carriers in the LTE mobile system is given by $\mathcal{K} = \{1, 2, \ldots, K\}$ and their corresponding offered prices are given by $\mathcal{P}^{\text{offered}} = \{p_1^{\text{offered}}, p_2^{\text{offered}}, \ldots, p_K^{\text{offered}}\}$. The jth user set of all in range carriers $\mathcal{K}_j$ (i.e., $\mathcal{K}_j = \{1, 2, \ldots, K_j\}$) corresponding offered prices are given by $\mathcal{P}^j = \{p_1^j, p_2^j, \ldots, p_{K_j}^j\}$.

The jth user all in range carriers $\mathcal{K}_j$ are arranged based on their offered prices as follows:

$$l_1^j = \arg\min_{\mathcal{K}_j} \{p_1^j, p_2^j, \ldots, p_{K_j}^j\}$$

$$\ell_2^j = \arg \min_{\mathcal{K}_j - \{\ell_1^j\}} \{p_1^j, p_2^j, \ldots, p_{K_j}^j\}$$

$$\vdots$$

$$\ell_{K_j}^j = \arg \min_{\mathcal{K}_j - \{\ell_1^j, \ldots, \ell_{K_j-1}^j\}} \{p_1^j, p_2^j, \ldots, p_{K_j}^j\}$$

where ℓ_1^j is the carrier with the lowest offered price and $\ell_{K_j}^j$ is the carrier with the highest offered price within the jth user set $\mathcal{K}_j$ of all in range carriers and $\mathcal{P}^j = \{p_1^j, p_2^j, \ldots, p_{K_j}^j\}$ is the set of the offered prices of all in range carriers for the jth user. The jth user sends an assignment of 1 to the ith eNodeB that is corresponding to eNodeB ℓ_1^j (i.e., the eNodeB with the least offered price among the jth user's all in range carriers). On the other hand, the jth user sends an assignment of 0 to each of the remaining eNodeBs in its range. Once the ith eNodeB receives an assignment of 1 from each UE in its coverage area, it starts allocating its resources to the M_i UEs in $\mathcal{M}_i$ such that the jth UE is allocated an optimal rate $r_i^{j,\text{opt}}$ from the ith eNodeB. Once the jth UE is allocated rate from its primary carrier ℓ_1^j, it then sends an assignment of 1 to the ith eNodeB that is corresponding to eNodeB ℓ_2^j and sends an assignment of 0 to each of the remaining eNodeBs in its range. The process continues until the jth UE sends an assignment of 1 to the ith eNodeB that is corresponding to eNodeB $\ell_{K_j}^j$ and receives its allocated rate from that eNodeB. The jth UE then calculates its aggregated final optimal rate r_j^{agg}.

3.2.2.2 Multi-Carrier RA Optimization Problem

Once the carriers offered prices are calculated as discussed in Sect. 3.2.2.1, each user j selects eNodeB ℓ_1^j to be its primary carrier and the remaining carriers in its range to be its secondary carriers. The eNodeB with the least offered price is the first one to start allocating its resources among all users in its coverage area. Each of the remaining eNodeBs then starts allocating its available resources after all the users in its coverage area are allocated rates from carriers in their range with lower offered prices. Eventually, each user j is allocated rates from all of the K_j carriers in its range. As discussed before, the ith carrier eNodeB starts allocating its resources among all users in its coverage area once it receives an assignment of 1 from each of the M_i users in $\mathcal{M}_i$. The rate allocated to the jth user from its ith carrier is given by $r_i^{j,\text{opt}}$.

The RA optimization problem for the ith carrier eNodeB in $\mathcal{K}$, such that the ith eNodeB received an assignment of 1 from each of the users under its coverage area, can be written as:

$$
\begin{aligned}
\max_{\mathbf{r_i}} \quad & \prod_{j=1}^{M_i} U_j\left(r_i^j + c_i^j\right) \\
\text{subject to} \quad & \sum_{j=1}^{M_i} r_i^j \le R_i, \\
& r_i^j \ge 0, \quad j = 1, 2, \ldots, M_i, \\
& c_i^j = \sum_{n=1, n\neq i}^{K} v_n^j r_n^{j,\text{opt}}, \\
& v_n^j = \begin{cases} 1, & \text{the } j\text{th UE} \in \mathcal{M}_n, \\ 0, & \text{the } j\text{th UE} \notin \mathcal{M}_n, \end{cases}
\end{aligned} \tag{3.13}
$$

where $\mathbf{r_i} = \{r_i^1, r_i^2, \ldots, r_i^{M_i}\}$, R_i is the ith eNodeB available resources, c_i^j is equivalent to the total rates allocated to the jth user by the carriers in its range with lower offered prices than the ith carrier offered price, v_n^j is equivalent to 1 if the jth UE $\in \mathcal{M}_n$ and is equivalent to 0 if the jth UE $\notin \mathcal{M}_n$, and $r_n^{j,\text{opt}}$ is the optimal rate allocated to the jth user by the nth eNodeB (i.e., the nth carrier $\in \mathcal{K}$). Once the jth user is allocated rate from all the carriers in its range, it then calculates its aggregated final optimal rate $r_j^{\text{agg}} = \sum_{i=1}^{K} v_i^j r_i^{j,\text{opt}}$.

Optimization problem (3.13) gives priority to the real-time application users and ensures that the minimum rate allocated to each user is c_i^j. Optimization problem (3.13) is a convex optimization problem and there exists a unique tractable global optimal solution [8]. The objective function in optimization problem (3.13) is equivalent to $\max_{\mathbf{r_i}} \sum_{j=1}^{M_i} \log U_j(r_i^j + c_i^j)$.

From optimization problem (3.13), we have the Lagrangian:

$$
L_i\left(r_i^j\right) = \left(\sum_{j=1}^{M_i} \log U_j\left(r_i^j + c_i^j\right)\right) - p_i\left(\sum_{j=1}^{M_i} r_i^j - R_i - z^i\right), \tag{3.14}
$$

where $z^i \ge 0$ is the slack variable and p_i is the Lagrange multiplier which is equivalent to the shadow price that corresponds to the ith carrier price per unit bandwidth for the M_i channels as in [8].

3.2.3 A Price Selective Centralized RA with CA Algorithm

In this section, we present our price selective centralized RA with CA algorithm. Each UE is allocated optimal rates from its all in range carriers and the final optimal rate allocated to each UE is the aggregated rate. Algorithms 5 and 6 are

Algorithm 5 The jth UE algorithm

Let $c_i^j = 0 \ \forall i \in \{1, 2, \ldots, K\}$
Send the UE application utility parameters k_j, a_j, and b_j to all in range eNodeBs
Receive offered prices that are equivalent to $\mathscr{P}^j = \{p_1^j, p_2^j, \ldots, p_{K_j}^j\}$ from all in range carriers eNodeBs
loop
  for $m \leftarrow 1$ to K_j **do**
    $l_m^j = \arg \min_{\mathscr{K}_j - \{l_1^j, \ldots, l_{m-1}^j\}} \{p_1^j, p_2^j, \ldots, p_{K_j}^j\}$ is carrier l_m^j for the jth UE
  end for
end loop
loop
  for $m \leftarrow 1$ to $K_j - 1$ **do**
    Send Flag assignment of 1 to the ith eNodeB and an assignment of 0 to the remaining carriers in $\mathscr{K}_j$ {eNodeB $i =$ eNodeB l_m^j}
    Send c_i^j to the ith eNodeB {eNodeB $i =$ eNodeB l_m^j}
    Receive the optimal rate $r_i^{j,opt}$ from the ith eNodeB {eNodeB $i =$ eNodeB l_m^j}
    Receive shadow price p_i from the ith eNodeB {eNodeB $i =$ eNodeB l_m^j}
    Send the optimal rate $r_i^{j,opt}$ to the ith eNodeB {the ith eNodeB corresponds to eNodeB l_{m+1}^j}
    Calculate new $c_i^j = \sum_{n=1, n \neq i}^{K} v_n^j r_n^{j,opt}$ for the ith eNodeB that corresponds to eNodeB l_{m+1}^j
  end for
end loop
Send c_i^j to the ith eNodeB {the ith carrier corresponds to carrier $l_{K_j}^j$}
Receive the optimal rate $r_i^{j,opt}$ from the ith eNodeB {the ith carrier corresponds to carrier $l_{K_j}^j$}
Receive shadow price p_i from the ith eNodeB {the ith carrier corresponds to carrier $l_{K_j}^j$}
Calculate the aggregated final optimal rate $r_j^{agg} = c_i^j + r_i^{j,opt}$ {the ith carrier corresponds to carrier $l_{K_j}^j$}

the UE and the eNodeB algorithms, respectively. The algorithm starts when each UE transmits its application parameters to all in range eNodeBs. Each eNodeB assigns initial values $w_{i,j}(0)$ to the users applications. Each eNodeB performs an internal iterative algorithm to calculate its offered price per unit bandwidth. In each iteration, the eNodeB checks the difference between the current value $w_{i,j}(n)$ and the previous one $w_{i,j}(n-1)$, if the difference is greater than a threshold δ, the shadow price $p_i^{\text{offered}}(n) = \frac{\sum_{j=1}^{M_i} w_{i,j}(n)}{R_i}$ is calculated by the eNodeB. Each eNodeB uses $p_i^{\text{offered}}(n)$ to calculate the rate $r_{i,j}(n)$ that is the solution of the optimization problem $r_{i,j}(n) = \arg\max_{r_{i,j}}(\log U_j(r_{i,j}) - p_i^{\text{offered}}(n) r_{i,j})$. The calculated rate is then used to calculate a new value $w_{i,j}(n)$ where $w_{i,j}(n) = p_i^{\text{offered}}(n) r_{i,j}(n)$. Each eNodeB checks the fluctuation condition as in [9] and calculates a new value $w_{i,j}(n)$. Once the difference between the current $w_{i,j}(n)$ and the previous one is less than δ for all UEs, the ith eNodeB sends its offered price p_i^{offered} to all UEs in its coverage area.

Once the jth UE receives the offered prices p_i^{offered} from all in range carriers, it sends an assignment of 1 to the ith eNodeB with the lowest offered price that is

Algorithm 6 The ith eNodeB algorithm

Let $w_{i,j}(0) = 0 \ \forall j \in \mathscr{M}_i$
Receive application utility parameters k_j, a_j, and b_j from all UEs under the coverage area of the ith eNodeB
loop
  while $|w_{i,j}(n) - w_{i,j}(n-1)| > \delta$ for any $j = \{1, \ldots, M_i\}$ where the jth UE under the coverage area of the ith eNodeB **do**
    Calculate $p_i^{\text{offered}}(n) = \frac{\sum_{j=1}^{M_i} w_{i,j}(n)}{R_i}$
    for $j \leftarrow 1$ to M_i **do**
      Solve $r_{i,j}(n) = \arg\max_{r_{i,j}} \Big(\log U_j(r_{i,j}) - p_i^{\text{offered}}(n) r_{i,j}(n) \Big)$
      Calculate new $w_{i,j}(n) = p_i^{\text{offered}}(n) r_{i,j}(n)$
      if $|w_{i,j}(n) - w_{i,j}(n-1)| > \Delta w$ **then**
        $w_{i,j}(n) = w_{i,j}(n-1) + \text{sign}(w_{i,j}(n) - w_{i,j}(n-1)) \Delta w(n)$
          $\{\Delta w(n) = l_1 e^{-\frac{n}{l_2}}\}$
      end if
    end for
  end while
  Send the ith eNodeB's shadow price $p_i^{\text{offered}} = p_i^{\text{offered}}(n) = \frac{\sum_{j=1}^{M_i} w_{i,j}(n)}{R_i}$ to all UEs in the eNodeB coverage area
end loop
if The ith eNodeB received Flag assignment of 1 from each UE (the jth UE where $j \in \mathscr{M}_i$) in its coverage area **then**
  loop
    Let $w_i^j(0) = 0 \ \forall j, j = \{1, \ldots, M_i\}$
    while $|w_i^j(n) - w_i^j(n-1)| > \delta$ for any $j = \{1, \ldots, M_i\}$ **do**
      Calculate $p_i(n) = \frac{\sum_{j=1}^{M_i} w_i^j(n)}{R_i}$
      for $j \leftarrow 1$ to M_i **do**
        Receive c_i^j value from the jth UE
        Solve $r_i^j(n) = \arg\max_{r_i^j} \Big(\log U_j(r_i^j + c_i^j) - p_i(n) r_i^j(n) \Big)$
        Calculate new $w_i^j(n) = p_i(n) r_i^j(n)$
        if $|w_i^j(n) - w_i^j(n-1)| > \Delta w$ **then**
          $w_i^j(n) = w_i^j(n-1) + \text{sign}(w_i^j(n) - w_i^j(n-1)) \Delta w(n)$
          $\{\Delta w(n) = l_1 e^{-\frac{n}{l_2}}\}$
        end if
      end for
    end while
    Send rate $r_i^{j,opt} = \frac{w_i^j(n)}{R_i}$ to all UEs in the eNodeB coverage area
    Send the shadow price $p_i = p_i(n)$ to all UEs in its coverage area
  end loop
end if

corresponding to eNodeB l_1^j and an assignment of 0 to the remaining eNodeBs in its range. The jth UE then receives its allocated rate $r_i^{j,\text{opt}}$ and shadow price p_i from that eNodeB. It then updates the c_i^j value and sends it to the ith eNodeB that is corresponding to eNodeB l_2^j, it also sends an assignment of 1 to that eNodeB and an assignment of 0 to the remaining eNodeBs in its range. The process continues until the jth UE receives its allocated rate $r_i^{j,\text{opt}}$ and shadow price p_{K_j}, and it then calculates its aggregated final optimal rate r_j^{agg}.

On the other hand, Once the ith eNodeB receives assignments of 1 from all UEs in its coverage area, it calculates the optimal rate $r_i^{j,\text{opt}}$ and shadow price p_i and sends them to each UE in its coverage area. The process continues until the eNodeB with the highest offered price receives assignment of 1 from all UEs in its coverage area, and it then sends each of these UEs its allocated optimal rate $r_i^{j,\text{opt}}$ and shadow price p_i.

3.2.4 Multi-Carrier Rate Allocation Numerical Results

Algorithms 5 and 6 were applied in C++ (Sect. 3.4.2) to different sigmoidal-like and logarithmic utility functions. The simulation results showed convergence to the global optimal rates. In this section, we present the simulation results for two carriers and nine UEs shown in Fig. 3.10. Three UEs {UE1, UE2, UE3} (first group) are under the coverage area of only Carrier 1 eNodeB, another three UEs {UE4, UE5, UE6} (second group) are joint users under the coverage area of both carrier 1 and carrier 2 eNodeBs, and three UEs {UE7, UE8, UE9} (third group) are under the coverage area of only carrier 2 eNodeB. UE1 and UE7 are running the same real-time application that is represented by a normalized sigmoidal-like utility function, that is expressed by Eq. (2.1), with $a = 5$, $b = 10$ which is an approximation to a step function at rate $r = 10$. UE2 and UE8 are running the same real-time

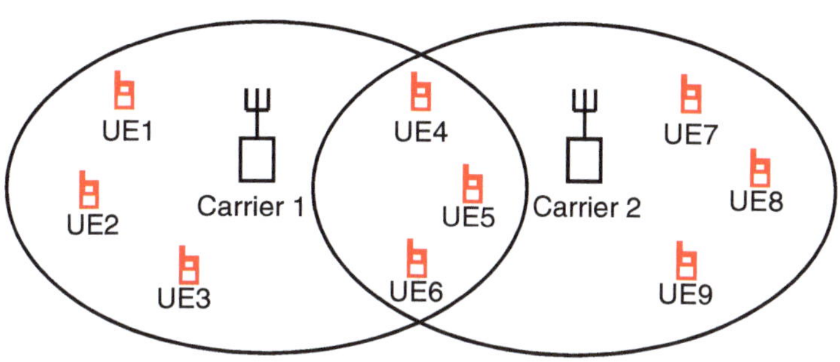

Fig. 3.10 System model with two carriers eNodeBs and three groups of users. UE1, UE2, and UE3 under the coverage area of only carrier 1. UE4, UE5, and UE6 under the coverage area of both carriers. UE7, UE8, and UE9 under the coverage area of only carrier 2

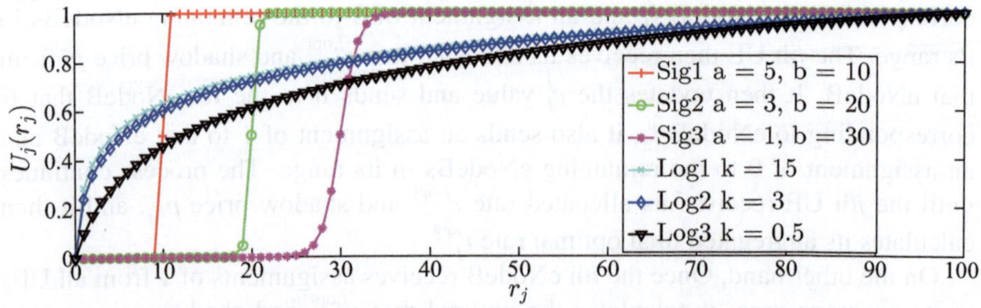

Fig. 3.11 The users utility functions $U_j(r_j)$. Sig1 represents UE1 and UE7 applications, Sig2 represents UE2 and UE8 applications, Log1 represents UE3 and UE9 applications, Log2 represents UE4 application, Log3 represents UE5 application, Sig3 represents UE6 application, and r_j is the rate allocated to the jth user from all in range eNodeBs

application that is represented by another sigmoidal-like utility function with $a = 3$ and $b = 20$. UE3 and UE9 are running the same delay-tolerant application that is represented by a logarithmic function with $k = 15$. The joint users UE4 and UE5 are running delay-tolerant applications that are represented by logarithmic functions with $k = 3$ and $k = 0.5$, respectively. The joint user UE6 is running real-time application that is represented by sigmoidal-like utility function with $a = 1$ and $b = 30$. Additionally, we use $r_{\max} = 100$ for all logarithmic functions, $l_1 = 5$ and $l_2 = 10$ in the fluctuation decay function of the algorithm, and $\delta = 10^{-3}$. The utility functions corresponding to the nine UEs applications are shown in Fig. 3.11.

3.2.4.1 Carrier Offered Price

In the following simulations, carrier 1 eNodeB available resources R_1 take values between 50 and 200 ($50 \leq R_1 \leq 200$) with step of 10, and carrier 2 eNodeB available resources are fixed $R_2 = 100$. In Fig. 3.12, we consider each carrier to be the primary carrier for all UEs under its coverage area and show that carrier 1 offered price p_1^{offered} is higher than carrier 2 offered price p_2^{offered} when $R_1 \leq R_2$ where $R_2 = 100$. On the other hand, Fig. 3.12 shows that $p_2^{\text{offered}} > p_1^{\text{offered}}$ when $R_2 < R_1 \leq 200$. This shows how the carrier's offered price depends on its available resources, the shadow price increases when the carrier's available resources decrease for a fixed number of users. As mentioned before, the joint users select the carrier with the lowest offered price to be their primary carrier. Therefore, in this case the joint users select carrier 2 to be their primary carrier and carrier 1 to be their secondary carrier when $R_1 \leq 100$ whereas they select carrier 1 to be their primary carrier and carrier 2 to be their secondary carrier when $100 < R_1 \leq 200$.

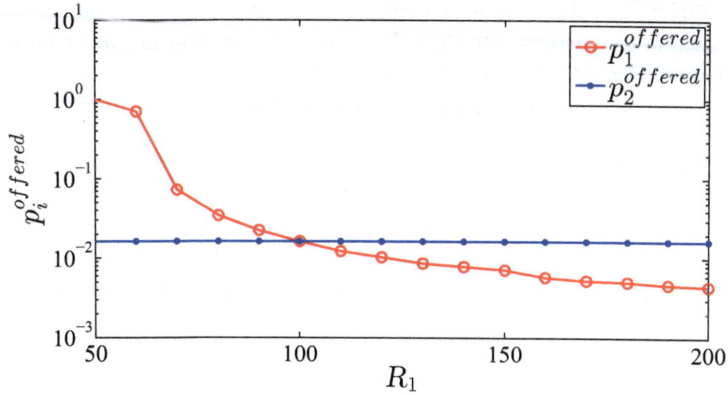

Fig. 3.12 Carrier 1 offered price p_1^{offered} for different values of R_1 and fixed number of users and carrier 2 offered price p_2^{offered} for $R_2 = 100$ assuming that each carrier is the primary carrier for all UEs under its coverage area

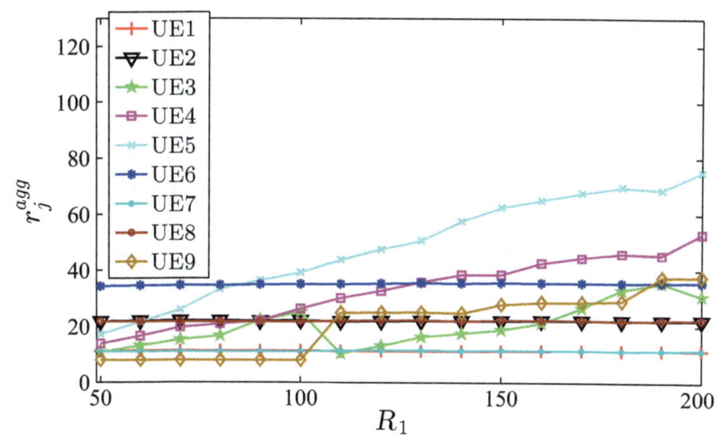

Fig. 3.13 The aggregated final optimal allocated rate r_j^{agg} for each user from its all in range carriers versus carrier 1 available resources $50 \leq R_1 \leq 200$ with carrier 2 available resources fixed at $R_2 = 100$

3.2.4.2 Aggregated Rates

In the following simulations, carrier 1 available resources R_1 take values between 50 and 200 ($50 \leq R_1 \leq 200$) with step of 10 and carrier 2 eNodeB available resources are fixed $R_2 = 100$. In Fig. 3.13, we show the aggregated final optimal rates for the nine users with different available resources R_1 of carrier 1. The final optimal rates r_j^{agg} for the first group of UEs are allocated to them by only carrier 1 as they are under the coverage area of only that carrier and do not have secondary carriers. Similarly, the final optimal rates r_j^{agg} for the third group of UEs are allocated to them by carrier 2 as they are under the coverage area of only that carrier and do not have secondary carriers. On the other hand, the second group of UEs are joint users and are allocated rates from both carriers. The joint users select their primary carrier l_1^j to be the carrier with the lowest shadow price $l_1^j = \arg\min_{\{1,2\}}\{p_1^{\text{offered}}, p_2^{\text{offered}}\}$ and the other carrier with a higher offered price to be their secondary carrier l_2^j. The aggregated final optimal rate allocated to each joint user is the aggregated rate of its primary carrier allocated rate and its secondary carrier allocated rate. Figure 3.13 shows that users running real-time applications are given priority over users running delay-tolerant applications and are allocated higher rates in the case of low carrier's available resources.

3.3 Summary and Conclusion

In this chapter, we introduced a novel RA with CA optimization framework in cellular networks. We considered mobile users with elastic or inelastic traffic and used utility functions to represent the applications running on the UEs. We presented an iterative decentralized rate allocation with CA algorithm to allocate both primary and secondary carriers resources optimally among users located under the coverage area of both carriers. We also presented a novel price selective centralized algorithm for allocating resources from different carriers optimally among users. Our price selective algorithm allows each user to select its primary and secondary carriers based on their offered prices in order to guarantee a minimum price for the aggregated final allocated rates. The centralized algorithm is performed mostly in the eNodeBs. Therefore, it requires less transmission overhead and less computations in the UEs. The proposed algorithms use proportional fairness approach to provide a minimum QoS to all users while giving priority to real-time applications. We analyzed the convergence of the algorithms with different carriers available resources and showed through simulations that our algorithms converge to optimal values.

3.4 C++ Codes

Some functions in the codes are listed in the Appendix.

3.4.1 First Code

```
#include<iostream>
#include<math.h>
#include <string>        //This is for string
#include <windows.h>     //This is for the console code

using namespace std;

float c_i[3];
const float lamda = 0.01;
float low = 0.001;
float weight_all[6][34];
float rate_all[6][34];
float rate_optimal_all[6][34];
float price_all[34];
float weight_all2[6][5];
float rate_all2[6][5];
float rate_optimal_all2[6][5];
float price_all2[5];
int counter = 0;
//---------------------------------------------------------//
//main
int main(){
    SetWindow(100,1000);         //This sets the size of the
        window to Width 100, Height 1000
    cout <<"prog is running...Done! \n"<<endl;
    rateCalc1();
         cout<<endl;
         cout<<endl;
         cout<<" price "<<endl;
    for(int j = 0; j<34; j++){

             cout<<price_all[j]<<" ";
    }
    for(int i = 0; i<6;i++){
         cout<<endl;
         cout<<endl;
         cout<<"user # "<<i<<endl;
         cout<<endl;
         cout<<" weight for user # "<<i<<endl;
         for(int j = 0; j<34; j++){
             cout<<weight_all[i][j]<<" ";
         }
         cout<<endl;
         cout<<" ri(n) for user # "<<i<<endl;
         for(int j = 0; j<34; j++){
         cout<<rate_all[i][j]<<" ";
         }
         cout<<endl;
         cout<<" optimal for user # "<<i<<endl;
```

```
            for(int j = 0; j<34; j++){
            cout<<rate_optimal_all[i][j]<< " ";
            }
        }
       //-------------------------------------//
       //Second Stage of the algorithm
       rateCalc2();
            cout<<endl;
            cout<<endl;
            cout<<" price "<<endl;
       for(int j = 0; j<5; j++){
            cout<<price_all2[j]<<" ";
       }
       for(int i = 0; i<6;i++){
            cout<<endl;
            cout<<endl;
            cout<<"user # "<<i<<endl;
            cout<<endl;
            cout<<" weight for user # "<<i<<endl;
            for(int j = 0; j<5; j++){
                cout<<weight_all2[i][j]<<" ";

            }
            cout<<endl;
            cout<<" ri(n) for user # "<<i<<endl;
            for(int j = 0; j<5; j++){
                cout<<rate_all2[i][j]<<" ";
            }
            cout<<endl;
            cout<<" optimal rate for user # "<<i<<endl;
            for(int j = 0; j<5; j++){
                cout<<rate_optimal_all2[i][j]<< " ";
            }

        }

        system("PAUSE");
       return 0;
}
//-----------------------------------------------------//
void rateCalc1(void){
    float weight[6] ={3.5,3.5,3.5,3.5,3.5,3.5};
    float prev_Weight[6] = {0,0,0,0,0,0};
    float p_n;
    float p_prev = 0;
    float k_i[3] = {15,3,.5}; //parameters for the logarithmic utility functions. One for each user
    float a_i[3] = {5,3,1}; //Parameters for the sigmoidal-like utility functions
    float b_i[3] = {10,20,30}; //Parameters for the sigmoidal like utility functions
    float opt1[6] ={0,0,0,0,0,0};//The constant opt is zero for each user since the calculations are made for stage 1 of the algorithm ( no aggregated rates yet).
```

```
    float opt2[6] ={0,0,0,0,0,0};
    int N_users = 6;
    float R = 70;
    int counter=0;


    p_n = shadowPrice(weight,R,N_users);
    cout<< p_n <<endl;
    float r_i_optimal[6];
    float rateArray[6];

    //Checking whether the difference between the current bid and the previous one is less than a threshold Δ for all users.
        while((abs(weight[0] - prev_Weight[0]) ≥ lamda) || (abs(weight[1] - prev_Weight[1]) ≥ lamda) || (abs(weight[2] - prev_Weight[2]) ≥ lamda) || (abs(weight[3] - prev_Weight[3]) ≥ lamda) || (abs(weight[4] - prev_Weight[4]) ≥ lamda) || (abs(weight[5] - prev_Weight[5]) ≥ lamda)){

        cout<<"\n" <<p_n<<" ->price #"<<counter <<endl; // prints out each price.
        price_all[counter] = p_n;

        //Printing  weight:
         for(int i=0; i<6; i++){
            cout<< weight[i] <<"    #" << i<<"   Weight"<< endl;
            weight_all[i][counter] = weight[i];
        }
        //Finding r_i_optimal
        for(int i = 0; i <6; i++){
            r_i_optimal[i] = weight[i]/p_n;
            cout<<r_i_optimal[i]<<"--" << i<<" ->optimal rate "<<endl;
            rate_optimal_all[i][counter] = r_i_optimal[i];
        }
        //Finding the ideal rates for users with sigmoidal-like utility functions
        for(int j = 0; j<3; j++){
            rateArray[j] = get_roots1(a_i[j],b_i[j],p_n,opt1[j],opt2[j],R);
            cout<<  rateArray[j]<<" #"<<j<<"     *** rate"<< endl;
            rate_all[j][counter] = rateArray[j];
        }
        //Finding the ideal rates for users with logarithmic utility functions
        for(int j=3; j<6; j++){
            rateArray[j] = get_roots2(k_i[j-3],p_n,opt1[j],opt2[j],R);
            cout<<  rateArray[j]<<" #"<<j<<"     *** rate"<< endl;
```

```
                rate_all[j][counter] = rateArray[j];
            }

            //Giving values to Weight:
            for(int k = 0; k<6; k++){
                prev_Weight[k] = weight[k];
            }
            //Finding new weights
            for(int k = 0; k<6; k++){
                weight[k] = p_n * rateArray[k];
            }

            p_prev = p_n;
            //Finding new shadow price based on the new weights
            p_n = shadowPrice(weight,R,N_users);
            cout<<p_n<<"--p_n_2"<<endl;
            counter++;
        }

            //Finding r_i_optimal for the last iteration
            cout<<"\n" <<p_n<<" ->price #"<<counter <<endl; // prints out each price.
            price_all[counter] = p_n;
            for(int i=0; i<6; i++){
                cout<< weight[i] <<"    #" << i<<"   Weight"<< endl;
                weight_all[i][counter] = weight[i];
            }
            for(int i = 0; i <6; i++){
                r_i_optimal[i] = weight[i]/p_n;
                cout<<r_i_optimal[i]<<"--" << i<<" ->optimal rate "<<endl;
                rate_optimal_all[i][counter] = r_i_optimal[i];
            }
            //Finding the ideal rate for the last iteration
            for(int j = 0; j<3; j++){
                rateArray[j] = get_roots1(a_i[j],b_i[j],p_n,opt1[j],opt2[j],R);
                cout<<  rateArray[j]<<" #"<<j<<"     *** rate"<< endl;
                rate_all[j][counter] = rateArray[j];
            }
            for(int j=3; j<6; j++){
                rateArray[j] = get_roots2(k_i[j-3],p_n,opt1[j],opt2[j],R);
                cout<<  rateArray[j]<<" #"<<j<<"     *** rate"<< endl;
                rate_all[j][counter] = rateArray[j];
            }
}

//-------------------------------------------------------//

void rateCalc2(void){
```

```
    float weight[6] = {0.25,0.25,0.25,0.25,0.25,0.25};
    float prev_Weight[6] = {0,0,0,0,0,0};
    float p_n;
    float p_prev = 0;
    float k_i[3] = {15,3,.5}; //parameters for the logarithmic utility functions. One for each user
    float a_i[3] = {5,3,1};   //Parameters for the sigmoidal-like utility functions
    float b_i[3] = {10,20,30}; //Parameters for the sigmoidal like utility functions
    float opt1[6] = {10.6393,20.8845,31.4117,1.53906,2.18847,3.26511}; // The constant opt is equal to the optimal rate given by the primary carrier in stage 1 of the Algorithm
    float opt2[6] = {0,0,0,0,0,0};
    int N_users = 6;
    float R = 50;
    int counter=0;


    p_n = shadowPrice(weight,R,N_users);
    cout<< p_n <<endl;
    float r_i_optimal[6];
    float rateArray[6];
    //Checking whether the difference between the current bid and the previous one is less than a threshold Δ for all users.
        while((abs(weight[0] - prev_Weight[0]) ≥ lamda) || (abs(weight[1] - prev_Weight[1]) ≥ lamda) || (abs(weight[2] - prev_Weight[2]) ≥ lamda) || (abs(weight[3] - prev_Weight[3]) ≥ lamda) || (abs(weight[4] - prev_Weight[4]) ≥ lamda) || (abs(weight[5] - prev_Weight[5]) ≥ lamda)){

        cout<<"\n" <<p_n<<" ->price #"<<counter <<endl; // prints out each price.
        price_all2[counter] = p_n;

        //Printing  weight:
        for(int i=0; i<6; i++){
            cout<< weight[i] <<"    #" << i<<"   Weight"<<endl;
            weight_all2[i][counter] = weight[i];
        }

        //Finding r_i_optimal
        for(int i = 0; i <6; i++){
            r_i_optimal[i] = weight[i]/p_n;
            cout<<r_i_optimal[i]<<"--" << i<<" ->optimal rate "<<endl;
            rate_optimal_all2[i][counter] = r_i_optimal[i];
        }
        //Finding the ideal rates for users with sigmoidal-like utility functions
```

```
        for(int j = 0; j<3; j++){
            rateArray[j] = get_roots1(a_i[j],b_i[j],p_n,opt1[
                j],opt2[j],R);
            cout<<  rateArray[j]<<" #"<<j<<"          *** rate"<<
                endl;
            rate_all2[j][counter] = rateArray[j];
        }
        //Finding the ideal rates for users with logarithmic
            utility functions
        for(int j=3; j<6; j++){
            rateArray[j] = get_roots2(k_i[j-3],p_n,opt1[j],
                opt2[j],R);
            cout<<  rateArray[j]<<" #"<<j<<"          *** rate"<<
                endl;
            rate_all2[j][counter] = rateArray[j];
        }

        //Giving values to Weight:
        for(int k = 0; k<6; k++){
            prev_Weight[k] = weight[k];
        }
       //Finding new weights
        for(int k = 0; k<6; k++){
            weight[k] = p_n * rateArray[k];
        }

        p_prev = p_n;
        //Finding new shadow price based on the new weights
        p_n = shadowPrice(weight,R,N_users);
        cout<<p_n<<"--p_n_2"<<endl;
        counter++;
    }

       //Finding r_i_optimal
        cout<<"\n" <<p_n<<" ->price #"<<counter <<endl; //
            prints out each price.
        price_all2[counter] = p_n;
          for(int i=0; i<6; i++){
            cout<< weight[i] <<"    #" << i<<"   Weight"<<
                endl;
            weight_all2[i][counter] = weight[i];
        }
        for(int i = 0; i <6; i++){
            r_i_optimal[i] = weight[i]/p_n;
            cout<<r_i_optimal[i]<<"--" << i<<" ->optimal rate
                "<<endl;
            rate_optimal_all2[i][counter] = r_i_optimal[i];
        }
        //Finding the ideal rate for the last iteration
        for(int j = 0; j<3; j++){
            rateArray[j] = get_roots1(a_i[j],b_i[j],p_n,opt1[
                j],opt2[j],R);
            cout<<  rateArray[j]<<" #"<<j<<"          *** rate"<<
                endl;
```

```
            rate_all2[j][counter] = rateArray[j];
        }
        for(int j=3; j<6; j++){
            rateArray[j] = get_roots2(k_i[j-3],p_n,opt1[j],
                opt2[j],R);
            cout<< rateArray[j]<<" #"<<j<<"      *** rate"<<
                endl;
            rate_all2[j][counter] = rateArray[j];
        }
}
```

3.4.2 Second Code

```
#include<iostream>
#include<math.h>
#include <string>        //This is for string
#include <windows.h>     //This is for the console code

using namespace std;

float c_i[3];
const float lamda = 0.001;
float weight_all[8][18];
float rate_all[8][18];
float rate_optimal_all[8][18];
float price_all[18];
int timer[18];
float Dif_R_Cont[6][18];
float Price_Matrix[18];
float Dif_R_Cont2[6][18];
float Price_Matrix22[18];
//----------------------------------//
//main
int main(){
    SetWindow(100,10000);        //This sets the size of the
        window to Width 100, Height 1000
    cout <<"prog is running...Done! \n"<<endl;
    float Dif_R_Cont1[6][18];
    float Dif_R_Cont22[6][1];
    int timer
        [18]={0,1,2,3,4,5,6,7,8,9,10,11,12,13,14,15,16,17};
    float rrate[18];
    for(int j = 0; j<18; j++){
        rrate[j]=((j+2)*10)+10;
        cout<<endl;
        cout<<"R = "<<rrate<<endl;
        cout<<endl;
    rateCalc1(rrate[j],timer[j]);
```

```
    }
        cout<<endl;
        cout<<endl;
        cout<<"Cont optimal Matrix "<<endl;
        cout<<endl;
        for(int i=0;i<6;i++){
        for(int j=0;j<18;j++){
            cout<<Dif_R_Cont[i][j]<<" "; // display the current element out of the array
            float value = Dif_R_Cont[i][j];
            Dif_R_Cont1[i][j] = value;
        }
    cout<<endl;
        }
    float Price_Matrix1[18];
         cout<<endl;
         cout<<endl;
         cout<<"Price Matrix "<<endl;
         cout<<endl;
        for(int j=0;j<18;j++){
            cout<<Price_Matrix[j]<<" ";  // display the current element out of the array
        }
    cout<<endl;  // when the inner loop is done, go to a new line

    float rrate2={100};
    int timer2={0};
         cout<<endl;
         cout<<"R2 = "<<rrate2<<endl;
         cout<<endl;
    rateCalc1(rrate2,timer2);
        cout<<endl;
        cout<<endl;
        cout<<"Cont optimal Matrix for carrier 2 "<<endl;
        cout<<endl;
        cout<<endl;
        for(int i=0;i<6;i++){
        for(int j=0;j<1;j++){
            cout<<Dif_R_Cont[i][j]<<" ";  // display the current element out of the array
            float value = Dif_R_Cont[i][j];
            Dif_R_Cont22[i][j] = value;
        }
        cout<<endl;  // when the inner loop is done, go to a new line
        }
         cout<<endl;
         cout<<endl;
         cout<<"Price Matrix "<<endl;
         cout<<endl;
          cout<<Price_Matrix[0]<<" ";  // display the current element out of the array
    float Price_Matrix2 = Price_Matrix[0];
```

```
    cout<<endl;  // when the inner loop is done, go to a new
        line
    float value0[6];
    float optimal_case[9][18];
    float optimal_price_case[3][18];
    float value00[6];
    float optimal_case2[9][18];

    cout<<"Final Optimal rates for the lowest price carrier
        cell"<<endl;
    cout<<endl;
    cout<<"1     "<<Dif_R_Cont22[3][0]<<endl;
    cout<<"2     "<<Dif_R_Cont22[4][0]<<endl;
    cout<<"3     "<<Dif_R_Cont22[5][0]<<endl;
    cout<<"4     "<<rrate2<<endl;
    cout<<endl;

    for(int j=0;j<8;j++){
        float rrate=((j+2)*10)+10;
         rateCalc2(rrate,timer[j],Dif_R_Cont22[3][0],
             Dif_R_Cont22[4][0],Dif_R_Cont22[5][0],rrate2);
        for(int i=0;i<3;i++){
         optimal_case[i][j]= Dif_R_Cont2[i][j];
        }
        for(int i=3;i<6;i++){
         optimal_case[i][j]= Dif_R_Cont2[i][j]+Dif_R_Cont22[i
             ][0];
        }
        for(int i=6;i<9;i++){
         optimal_case[i][j]= Dif_R_Cont22[i-6][0];
        }
        float val=Price_Matrix22[j];
        optimal_price_case[0][j]=val;
        optimal_price_case[1][j]=val+Price_Matrix2;
        optimal_price_case[2][j]=Price_Matrix2;
    }

    for(int j=8;j<18;j++){
    float rrate2={100};
    int timer2=j;
         rateCalc2(rrate2,timer2,Dif_R_Cont1[3][j],
             Dif_R_Cont1[4][j],Dif_R_Cont1[5][j],rrate[j]);
        for(int i=0;i<3;i++){
         optimal_case[i][j]= Dif_R_Cont1[i][j];
        }
        for(int i=3;i<6;i++){
         optimal_case[i][j]= Dif_R_Cont2[i][j]+Dif_R_Cont1[i
             ][j];
        }
        for(int i=6;i<9;i++){
         optimal_case[i][j]= Dif_R_Cont2[i-6][j];
        }
        float val=Price_Matrix[j];
```

```
        optimal_price_case[0][j]=val;
        optimal_price_case[1][j]=val+Price_Matrix22[j];
        optimal_price_case[2][j]=Price_Matrix22[j];
    }

     cout<<endl;
     cout<<endl;
     cout<<"Final optimal rates Matrix "<<endl;
     cout<<endl;
    for(int i=0;i<9;i++){
    for(int j=0;j<18;j++){
        cout<< optimal_case[i][j]<<" "; // display the
            current element out of the array
    }
    cout<<endl;  // when the inner loop is done, go to a
        new line
    }
     cout<<endl;
     cout<<endl;
     cout<<"Final Price Matrix "<<endl;
     cout<<endl;
    for(int i=0;i<3;i++) {
    for(int j=0;j<18;j++) {
        cout<<optimal_price_case[i][j]<<" ";  // display
            the current element out of the array
    }
    cout<<endl; // when the inner loop is done, go to a
        new line
    }
    cout<<endl;

  system("PAUSE");
 return 0;
}
//-------------------------------------------------------//
void rateCalc1(float Rate,int time){
    float weight1[] = {3.5,3.5,3.5,3.5,3.5,3.5};
    float prev_Weight1[6] = {0,0,0,0,0,0};
    float weight2[6] = {0,0,0,0,0,0};
    float p_n_1=0;
    float p_prev_1 = 0;
    float k_i[6] = {15,3,0.5};
    float a_i[6] = {5,3,1};
    float b_i[6] = {10,20,30};
    float rateArray1[6] = {0,0,0,0,0,0};
    float opt1[6] = {0,0,0,0,0,0};
    float opt2[6] = {0,0,0,0,0,0};
    float rate = Rate;
    int counter = 0;
    int N_users = 6;
    float r_i_optimal_1[6];

    p_n_1 = shadowPrice(weight1,rate,N_users);

```

```
        while (((abs(weight1[0] - prev_Weight1[0]) ≥ lamda) || (
            abs(weight1[1] - prev_Weight1[1]) ≥ lamda) || (abs(
            weight1[2] - prev_Weight1[2]) ≥ lamda) || (abs(weight1
            [3] - prev_Weight1[3]) ≥ lamda) || (abs(weight1[4] -
            prev_Weight1[4]) ≥ lamda) || (abs(weight1[5] -
            prev_Weight1[5]) ≥ lamda))){

            //Finding ri
            for(int i = 0; i <6; i++){
                r_i_optimal_1[i] = weight1[i]/p_n_1;
            }
            //Finding ri_optimal
            for(int j = 0; j<2; j++){
                rateArray1[j] = get_roots1(a_i[j],b_i[j],p_n_1,
                    opt1[j],opt2[j],rate);
            }
            for(int j = 2; j<3; j++){
                rateArray1[j] = get_roots2(k_i[j-2],p_n_1,opt1[j
                    ],opt2[j],rate);
            }
            for(int j=3; j<5; j++){
                rateArray1[j] = get_roots2(k_i[j-2],p_n_1,opt1[j
                    ],opt2[j],rate);
            }
            for(int j = 5; j<6; j++){
                rateArray1[j] = get_roots1(a_i[j-3],b_i[j-3],
                    p_n_1,opt1[j],opt2[j],rate);
            }
            //Giving values to Weight:
            for(int k = 0; k<6; k++){
            prev_Weight1[k] = weight1[k];
            }
            for(int k = 0; k<6; k++){
            weight2[k] = p_n_1 * rateArray1[k];
            }
        /////////Fluctuation Decay Condition/////////
        for(int k = 0; k<6; k++){
                float exp = 2.718281828459;
                float Del_W= 5*pow(exp, (-counter/10));
                if (abs(weight2[k]-prev_Weight1[k])>Del_W){ //
                    Check fluctuation condition
                weight1[k] = prev_Weight1[k]+(((weight2[k]-
                    prev_Weight1[k])/abs(weight2[k]-prev_Weight1[k
                    ]))*Del_W);  //Calculate new bid
                }
                else
                weight1[k]=weight2[k];
        }
            p_prev_1 = p_n_1;
            p_n_1 = shadowPrice(weight1,rate,N_users);
            counter++;
        }
        for(int j = 0; j<6; j++){
                float Num1=r_i_optimal_1[j];
```

```
            Dif_R_Cont[j][time]=Num1;
        }
        float Num2=p_n_1;
        Price_Matrix[time]=Num2;
    }
    //---------------------------------------------------//
    void rateCalc2(float Rate,int time,float optt1,float optt2,
        float optt3,float R22){
        float weight1[] = {3.5,3.5,3.5,3.5,3.5,3.5};
        float prev_Weight1[6] = {0,0,0,0,0,0};
        float weight2[6] = {0,0,0,0,0,0};
        float p_n_1 = 0;
        float p_prev_1 = 0;
        float k_i[6] ={15,3,0.5};
        float a_i[6] = {5,3,1};
        float b_i[6] = {10,20,30};
        float rateArray1[6] = {0,0,0,0,0,0};
        float opt1[6] = {0,0,0,optt1,optt2,optt3};
        float opt2[6] = {0,0,0,0,0,0};
        float rate = Rate;
        int counter = 0;
        int N_users = 6;
        float r_i_optimal_1[6];

        p_n_1 = shadowPrice(weight1,rate,N_users);

        while (((abs(weight1[0] - prev_Weight1[0]) ≥ lamda) || (
            abs(weight1[1] - prev_Weight1[1]) ≥ lamda) || (abs(
            weight1[2] - prev_Weight1[2]) ≥ lamda) || (abs(weight1
            [3] - prev_Weight1[3]) ≥ lamda) || (abs(weight1[4] -
            prev_Weight1[4]) ≥ lamda) || (abs(weight1[5] -
            prev_Weight1[5]) ≥ lamda))){

            //Finding ri
            for(int i = 0; i <6; i++){
                r_i_optimal_1[i] = weight1[i]/p_n_1;
            }
            //Finding ri_optimal
            for(int j = 0; j<2; j++){
                rateArray1[j] = get_roots1(a_i[j],b_i[j],p_n_1,
                    opt1[j],opt2[j],rate);
            }
            for(int j = 2; j<3; j++){
                rateArray1[j] = get_roots2(k_i[j-2],p_n_1,opt1[j
                    ],opt2[j],rate);
            }
            for(int j=3; j<5; j++){
                rateArray1[j] = get_roots2(k_i[j-2],p_n_1,opt1[j
                    ],opt2[j],rate);
            }
            for(int j = 5; j<6; j++){
                rateArray1[j] = get_roots1(a_i[j-3],b_i[j-3],
                    p_n_1,opt1[j],opt2[j],rate);
            }
```

```

        // Giving values to Weight:
        for(int k = 0; k<6; k++){

            prev_Weight1[k]  = weight1[k];
        }

        for(int k = 0; k<6; k++){
            weight2[k]  = p_n_1 * rateArray1[k];
        }
    /////////Fluctuation Decay Condition/////////
    for(int k = 0; k<6; k++){
            float exp = 2.718281828459;
            float Del_W= 5*pow(exp, (-counter/10));
            if (abs(weight2[k]-prev_Weight1[k])>Del_W){ //
                Check fluctuation condition
            weight1[k]  = prev_Weight1[k]+(((weight2[k]  -
                prev_Weight1[k])/abs(weight2[k]-prev_Weight1[k
                ]))*Del_W);    //Calculate new bid
            }
            else
            weight1[k]=weight2[k];
    }
        p_prev_1 = p_n_1;
        p_n_1 = shadowPrice(weight1,rate,N_users);
        counter++;
    }
    for(int j = 0; j<6; j++){
            float Num1=r_i_optimal_1[j];
            Dif_R_Cont2[j][time]=Num1;
    }
    float Num2=p_n_1;
    Price_Matrix22[time]=Num2;
}
```

References

1. H. Shajaiah, A. Abdelhadi, C. Clancy, Utility proportional fairness resource allocation with carrier aggregation in 4G-LTE, in *IEEE Military Communications Conference (MILCOM)* (2013)
2. H. Shajaiah, A. Abdelhadi, C. Clancy, A price selective centralized algorithm for resource allocation with carrier aggregation in LTE cellular networks, in *2015 IEEE Wireless Communications and Networking Conference (WCNC)*, March 2015, pp. 813–818
3. Z. Shen, A. Papasakellariou, J. Montojo, D. Gerstenberger, F. Xu, Overview of 3GPP LTE-advanced carrier aggregation for 4G wireless communications. IEEE Commun. Mag. **50**, 122–130 (2012)
4. *Frequency Spectrum Wall Chart* (Commerce Dept., National Telecommunications and Information Administration, Office of Spectrum Management, 2016)
5. G. Yuan, X. Zhang, W. Wang, Y. Yang, Carrier aggregation for LTE-advanced mobile communication systems. IEEE Commun. Mag. **48**, 88–93 (2010)

6. F. Wilson, I. Wakeman, W. Smith, Quality of service parameters for commercial application of video telephony, in *Proceedings of the Human Factors in Telecommunications Symposium* (Darmstadt, 1993), pp. 139–148
7. G. Tychogiorgos, A. Gkelias, K.K. Leung, Utility-proportional fairness in wireless networks, in *PIMRC* (IEEE, New York, 2012), pp. 839–844
8. A. Abdelhadi, C. Clancy, A utility proportional fairness approach for resource allocation in 4G-LTE, in *IEEE International Conference on Computing, Networking, and Communications (ICNC), CNC Workshop* (2014)
9. A. Abdelhadi, C. Clancy, A robust optimal rate allocation algorithm and pricing policy for hybrid traffic in 4G-LTE, in *IEEE International Symposium on Personal, Indoor, and Mobile Radio Communications (PIMRC)* (2013)
10. Y. Wang, A. Abdelhadi, A QoS-based power allocation for cellular users with different modulations, in *2016 International Conference on Computing, Networking and Communications (ICNC)*, February 2016, pp. 1–5
11. A. Abdelhadi, H. Shajaiah, Optimal resource allocation for smart phones with multiple applications with MATLAB instructions. Technical Reports (2016)
12. A. Abdelhadi, H. Shajaiah, Application-aware resource allocation with carrier aggregation using MATLAB. Technical Reports (2016)
13. Z. Kbah, A. Abdelhadi, Resource allocation in cellular systems for applications with random parameters, in *2016 International Conference on Computing, Networking and Communications (ICNC)*, February 2016, pp. 1–5
14. A. Abdelhadi, H. Shajaiah, Optimal resource allocation for cellular networks with MATLAB instructions. CoRR, abs/1612.07862 (2016)
15. J.-W. Lee, R.R. Mazumdar, N.B. Shroff, Downlink power allocation for multi-class wireless systems. IEEE/ACM Trans. Netw. **13**, 854–867 (2005)
16. S. Shenker, Fundamental design issues for the future internet. IEEE J. Sel. Areas Commun. **13**(7), 1176–1188 (1995)
17. Y. Wang, A. Abdelhadi, T.C. Clancy, Optimal power allocation for LTE users with different modulations, in *2016 Annual IEEE Systems Conference (SysCon)*, April 2016, pp. 1–5
18. A. Kumar, A. Abdelhadi, T.C. Clancy, A delay efficient multiclass packet scheduler for heterogeneous M2M uplink, in *IEEE MILCOM* (2016)
19. A. Kumar, A. Abdelhadi, T.C. Clancy, An online delay efficient packet scheduler for M2M traffic in industrial automation, in *IEEE Systems Conference* (2016)
20. A. Kumar, A. Abdelhadi, T.C. Clancy, A delay optimal MAC and packet scheduler for heterogeneous M2M uplink. CoRR, abs/1606.06692 (2016)
21. A. Kumar, A. Abdelhadi, T.C. Clancy, A delay-optimal packet scheduler for M2M uplink, in *IEEE MILCOM* (2016)
22. M. Ghorbanzadeh, A. Abdelhadi, C. Clancy, Distributed resource allocation, in *Cellular Communications Systems in Congested Environments* (Springer, Cham, 2017), pp. 61–91
23. J.B. Taylor, *Principles of Microeconomics*. Microeconomics Series (Houghton Mifflin, Boston, 1998)
24. S. Boyd, L. Vandenberghe, *Introduction to Convex Optimization with Engineering Applications*. Convex Optimization, Course Reader for EE364 (1999)
25. R. Madan, S. Boyd, S. Lall, Fast algorithms for resource allocation in wireless cellular networks. IEEE/ACM Trans. Networking **18**, 973–984 (2010)

Chapter 4
Resource Allocation with User Discrimination for Spectrum Sharing

In this chapter, we focus on the problem of radio resource allocation with user discrimination for different scenarios in cellular networks. First, we present a resource allocation with user discrimination approach for spectrum sharing between public safety and commercial users. It is important to have a common technical standard for commercial and public safety users as it provides advantages for both. The public safety systems market is much smaller than the commercial cellular market which makes it unable to attract the level of investment that goes in to commercial cellular networks and this makes a common technical standards for both the best solution. The public safety community gains access to the technical advantages provided by the commercial cellular networks whereas the commercial cellular community gains enhancement in their systems and makes it more attractive to consumers. The National Public Safety Telecommunications Council (NPSTC) and other organizations recognized the desirability of having an inter operable national standard for a next generation public safety network with broadband capabilities. The USA has reserved spectrum in the 700 MHz band for an LTE based public safety network. The current public safety standards support medium speed data which drives the need of new technology to add true mobile broadband capabilities and makes LTE the baseline technology for next generation broadband public safety networks.

Then, we provide a resource allocation with user discrimination optimization framework in cellular networks for different types of users running multiple applications simultaneously. Mobile users are now running multiple applications simultaneously on their smart phones. Operators are moving from single-service

---

The content in this chapter is reproduced with permission after modifications (License numbers 4078230417133 and 4078230144086). For the original article please refer to [1, 2].

H. Shajaiah et al., *Resource Allocation with Carrier Aggregation in Cellular Networks*, DOI 10.1007/978-3-319-60540-1_4

to multi-service and new services such as multimedia telephony and mobile-TV are now provided. In addition, different users subscribing for the same service may receive different treatment from the network providers [3–8] because of the subscriber differentiation provided by the service providers.

In addition, we present an efficient resource allocation with user discrimination framework for 5G Wireless Systems to allocate multiple carriers resources among users with elastic and inelastic traffic. As 5G systems' expected capabilities have started to take shape, CA is expected to be supported by 5G. Therefore, CA needs to be taken into consideration when designing 5G systems. Beside CA capability, 5G wireless network promises to handle diverse QoS requirements of multiple applications since different applications require different application's performance [9–11]. Furthermore, certain types of users may require to be given priority when allocating the network resources (i.e., public safety users) which needs to be taken into consideration when designing the resource allocation framework.

4.1 Spectrum Sharing Between Public Safety and Commercial Users in Cellular Networks

In this section, we propose a spectrum sharing approach between two groups of users, public safety and commercial users. We focus on finding an optimal solution for the resource allocation problem for the two groups of users running applications that are presented by logarithmic utility functions or sigmoidal-like utility functions. These utility functions are concave and non-concave utility functions, respectively. The optimization problem allocates part of the bandwidth from one eNodeB to each user subscribing for a mobile service taking into consideration that each user is getting a minimum QoS. In addition, the public safety users in emergency mode are given priority over the commercial users and within each group the non-concave functions that are approximated by sigmoidal-like functions and representing real-time applications are given priority over the concave functions approximated by logarithmic functions and presenting delay-tolerant applications. In our system model, each public safety subscriber has an assigned application target rate that varies based on the application type and assigned to the public safety subscriber by the network.

Our resource allocation algorithm first allocates the application target rate to each public safety UE when that UE is in emergency mode. It then allocates the remaining resources among the commercial UEs subscribing for resources. The contributions in this section are summarized as:

- We present a resource allocation optimization problem to allocate the eNodeB resources optimally among public safety and commercial users. The eNodeB and the UE collaborate to allocate an optimal rate to each UE with priority given to public safety users. Within the same group of users, a priority is given to real-time applications presented by sigmoidal-like utility functions.

- We show that each of our two cases resource allocation (RA) optimization problems has a unique tractable global optimal solution.

4.1.1 Spectrum Sharing Problem Formulation

We consider a single cell 4G-LTE mobile system with a single eNodeB, N commercial UEs, and M public safety UEs. The user i is allocated certain rate r_i based on the type of application the UE is running. Each user is assigned a utility function $U_i(r_i)$ based on the application running on the UE and whether it is a commercial or public safety user. Our goal is to determine the optimal bandwidth that needs to be allocated to each user by the eNodeB.

Utility functions $U_i(r_i)$ are used to represent the applications running on the UEs [11–20]. Logarithmic utility functions expressed by Eq. (2.2) and sigmoidal-like utility functions expressed by Eq. (2.1) are used to represent delay-tolerant and real-time applications, respectively, as in [11, 13, 21–27]. The basic formulation of the resource allocation problem is given by the following optimization problem:

$$\begin{aligned} \max_{\mathbf{r}} \quad & \prod_{i=1}^{M} U_i(r_{i,s}) \prod_{j=1}^{N} U_j(r_{j,c}) \\ \text{subject to} \quad & \sum_{i=1}^{M} r_{i,s} + \sum_{j=1}^{N} r_{j,c} \leq R, \\ & r_{i,s} \geq r_{i,s}^{t}, \quad i = 1,2,\ldots,M \\ & r_{j,c} \geq 0, \quad j = 1,2,\ldots,N \end{aligned} \tag{4.1}$$

where R is the maximum achievable rate of the eNodeB, $\mathbf{r} = \{r_{1,s}, \ldots, r_{M,s}, r_{1,c}, \ldots, r_{N,c}\}$ where $r_{i,s}$ is the rate for public safety user i, $r_{j,c}$ is the rate for commercial user j, $r_{i,s}^t$ is the application target rate for public safety user i which is the minimum rate that the user wants to achieve, M and N are the numbers of the public safety and commercial UEs, respectively. The resource allocation objective function maximizes the product of users utilities (system utility) when allocating resources to each user. Therefore, it provides a proportional fairness among utilities. Public safety users that are running real-time applications are given the priority when allocating resources by the eNodeB. The next priority is given to the elastic traffic running by public safety users. Once each public safety user satisfies its application target rate the eNodeB starts allocating resources to commercial users giving priority to users running real-time applications. We assume that the public safety users are in an emergency mode, therefore these users are given higher priority over commercial users. The optimization problem (4.1) has a unique tractable global optimal solution [14] that will be discussed in the next section.

We used utility proportional fairness model because non-zero rate allocation is guaranteed to all users. So it is impossible to set a user's allocation to zero without setting the efficiency of the network to zero. Because this resource allocation strategy does not disenfranchise any given user, it will be considered as an appropriate fairness model for this problem.

4.1.2 Spectrum Sharing Optimization

The resource allocation for public safety and commercial users is divided into two cases. The first case is when the maximum available resources R for the eNodeB is less than the sum of the total application target rates of the public safety UEs subscribing for a service from that eNodeB and the second case is when R is greater than that total. The two cases are two different optimization problems that will be solved by our proposed algorithm to obtain the optimal rate for each UE.

4.1.2.1 The First Case RA Optimization Problem When $\sum_{i=1}^{M} r_{i,s}^{t} \geq R$

As mentioned before the first case optimization problem is applied in the case of $\sum_{i=1}^{M} r_{i,s}^{t} \geq R$. In this case the eNodeB only allocates resources to the public safety users because they are considered more important and the eNodeB's available resources doesn't exceed their need. Commercial users will not be given any of the eNodeB resources in this case. This optimization problem can be written as:

$$
\begin{aligned}
\max_{\mathbf{r}} \quad & \prod_{i=1}^{M} U_i(r_{i,s}) \\
\text{subject to} \quad & \sum_{i=1}^{M} r_{i,s} \leq R, \\
& 0 \leq r_{i,s} \leq r_{i,s}^{t}, \qquad i = 1, 2, \ldots, M
\end{aligned}
\tag{4.2}
$$

where U_i is the public safety ith utility function, $\mathbf{r} = \{r_{1,s}, \ldots, r_{M,s}\}$ and M is the number of public safety UEs in the coverage area of the eNodeB. The solution of the optimization problem (4.2) is the optimal solution when $\sum_{i=1}^{M} r_{i,s}^{t} \geq R$. This solution will guarantee that the public safety users are given priority when allocating the eNodeB resources. The optimal rate for each public safety UE is less than or equal to the application target rate for each public safety UE. Public safety users that are running real-time applications will be given priority over public safety users with elastic traffic.

The objective function in the optimization problem (4.2) is equivalent to $\max_{\mathbf{r}} \sum_{i=1}^{M} \log U_i(r_{i,s})$. The optimization problem (4.2) is a convex optimization problem [28–30] and there exists a unique tractable global optimal solution as shown in Theorem (III.1) [14]. This optimal solution gives each of the M users an optimal rate $r_{i,s}^{\text{opt}}$.

4.1.2.2 The Second Case RA Optimization Problem When $\sum_{i=1}^{M} r_{i,s}^{t} < R$

The second case optimization problem is applied in the case of $\sum_{i=1}^{M} r_{i,s}^{t} < R$. The eNodeB collaborates with the UEs to solve this optimization problem. The eNodeB allocates resources to both public safety and commercial users because its available resources exceed the minimum need of the public safety UEs expressed by the application target rates. As mentioned before, the eNodeB gives priority to the public safety users and within the public safety group the priority is given to the UEs running inelastic traffic. This optimization problem can be written as:

$$\begin{aligned}
\max_{\mathbf{r}} \quad & \prod_{i=1}^{M} U_i(r_{i,s}) \prod_{j=1}^{N} U_j(r_{j,c}) \\
\text{subject to} \quad & \sum_{i=1}^{M} r_{i,s} + \sum_{j=1}^{N} r_{j,c} \leq R, \\
& r_{i,s} \geq r_{i,s}^{t}, \quad i = 1, 2, \ldots, M \\
& r_{j,c} \geq 0, \quad j = 1, 2, \ldots, N.
\end{aligned} \tag{4.3}$$

This optimization problem is same as the one discussed in the problem formulation (Sect. 4.1.1). First, the eNodeB allocates the application target rate to each public safety UE. It then starts allocating its remaining resources both public safety and commercial UEs based on utility proportional fairness. The solution of the optimization problem (4.3) is the global optimal solution that gives an optimal rate $r_{i,s}^{\text{opt}}$ to each public safety UE and an optimal rate $r_{i,c}^{\text{opt}}$ to each commercial user UE.

Proposition 4.1.1 *The optimization problem (4.3) is a convex optimization problem and there exists a unique tractable global optimal solution.*

Proof We introduce a new parameter c_i where c_i is equivalent to the application target rate for the public safety UE whereas it is equivalent to 0 for a commercial UE, the optimization problem (4.3) can be rewritten as follows:

$$\max_{\mathbf{r}} \prod_{i=1}^{M+N} U_i(r_i + c_i)$$

$$\text{subject to} \quad \sum_{i=1}^{M+N} (r_i + c_i) \le R, \tag{4.4}$$

$$r_i \ge 0, \quad i = 1, 2, \ldots, M + N$$

$$c_i = \begin{cases} r^t_{i,s} & \text{if public safety UE} \\ 0 & \text{if commercial UE} \end{cases}$$

where R is the maximum achievable rate of the eNodeB, $\mathbf{r} = \{r_1, \ldots, r_M, r_{M+1}, \ldots, r_{M+N}\}$ where the first M rates are for the M public safety users and the last N rates are for the N commercial users, $U_i(r_i + c_i)$ is the UE utility function, this optimization problem guarantees an optimal rate that is at least equal to the application target rate for the public safety UE. The objective function in the optimization problem (4.4) can be written as $\sum_{i=1}^{M+N} \log U_i(r_i + c_i)$.

The utility function $U_i(r_i + c_i)$ for the UE is strictly concave or sigmoidal-like function as mentioned in Sect. 4.1.1. As shown in Theorem (III.1) [14], $\log U_i(r_i)$ is a strictly concave function for a strictly concave or sigmoidal-like utility function. It follows that the optimization problem (4.4) that is equivalent to (4.3) is convex. Therefore, there exists a tractable global optimal solution for the optimization problem (4.3).

4.1.3 Algorithm for Optimal RA with Spectrum Sharing

We propose an iterative algorithm to allocate the system resources optimally for the spectrum sharing system described above. The eNodeB and the UEs collaborate to allocate optimal rates for the public safety and commercial users subscribing for a mobile service. Algorithms 7 and 8 are the public safety UE and the commercial UE algorithms, respectively. Algorithm 9 is the eNodeB algorithm. The algorithm starts when each UE transmits an initial bid $w_i(1)$ to the eNodeB. Additionally, each public safety UE transmits its application target rate to the eNodeB. The eNodeB checks whether $\sum_{i=1}^{M} r^t_{i,s}$ is less or greater than R and sends a flag with this information to each UE. In the case of $\sum_{i=1}^{M} r^t_{i,s} \ge R$, the commercial UEs will not be allocated any of the resources and will not be sending any further bids to the eNodeB unless they receive a flag from the eNodeB with $\sum_{i=1}^{M} r^t_{i,s} < R$.

On the other hand, each public safety UE checks whether the difference between the current received bid and the previous one is less than a threshold δ, if so it exits. Otherwise, if the difference is greater than δ, eNodeB calculates the shadow price $p(n) = \frac{\sum_{i=1}^{M} w_i(n)}{R}$. The estimated $p(n)$ is then sent to the public safety UEs where it is used to calculate the rate $r_{i,s}(n)$ which is the solution of the optimization problem $r_{i,s}(n) = \arg\max_{r_{i,s}}(\log U_i(r_{i,s}) - p(n)r_{i,s})$. A new bid $w_i(n)$ is calculated using $r_i(n)$ where $w_i(n) = p(n)r_{i,s}(n)$. All public safety UEs send their new bids $w_i(n)$ to the eNodeB. The algorithm is finalized by the eNodeB. Each public safety UE then calculates its allocated rate $r_{i,s}^{\text{opt}} = \frac{w_i(n)}{p(n)}$.

In the case of $\sum_{i=1}^{M} r_{i,s}^{t} < R$, the eNodeB sends a flag with this information to each UE. Each public safety and commercial UE checks whether the difference between the current received bid and the previous one is less than a threshold δ, if so it exits. Otherwise, if the difference is greater than δ, eNodeB calculates the shadow price $p(n) = \frac{\sum_{i=1}^{M+N} w_i(n)}{R}$. The estimated $p(n)$ is then sent to the public safety and commercial UEs where it is used by the public safety UE to calculate the rate $r_{i,s}(n) = r_i + r_{i,s}^{\text{t}}$ which is the solution of the optimization problem $r_{i,s}(n) = \arg\max_{r_{i,s}}(\log U_i(r_i + c_i) - p(n)(r_i + c_i))$. A new bid $w_i(n)$ is calculated by the public safety UE using $r_i(n)$ where $w_i(n) = p(n)(r_i(n) + c_i)$. All public safety UEs send their new bids $w_i(n)$ to the eNodeB. On the other hand, the commercial UEs receive $p(n)$ and use it to calculate the rate $r_{i,c}(n)$ which is the solution of the optimization problem $r_{i,c}(n) = \arg\max_{r_{i,c}}(\log U_i(r_{i,c}) - p(n)r_{i,c})$. A new bid $w_i(n)$ is calculated by the commercial UE using $r_{i,c}(n)$ where $w_i(n) = p(n)r_i(n)$. All public safety UEs send their new bids $w_i(n)$ to the eNodeB. The algorithm is finalized by the eNodeB. Each public safety UE then calculates its allocated rate $r_{i,s}^{\text{opt}} = \frac{w_i(n)}{p(n)}$ and each commercial UE calculates its allocated rate $r_{i,c}^{\text{opt}} = \frac{w_i(n)}{p(n)}$.

4.1.4 Spectrum Sharing Simulation Results

We consider one eNodeB with four public safety UEs and another four commercial UEs in its coverage area. We use multiple sigmoidal-like and logarithmic utility functions in our simulations and present two cases, one when the eNodeB resources R is less than the total application target rates of the public safety UEs and the other when R is greater than that total. We applied Algorithm 7–9 in C++ (Sect. 4.5.1) to the sigmoidal-like and logarithmic utility functions. The simulation results showed convergence to the optimal global point in both cases. We present the simulation results for eight utility functions that correspond to public safety and commercial UEs running real-time applications or delay-tolerant applications. We use two normalized utility functions expressed in Eq. (2.1) with different parameters a and b for each utility function, $a = 3$, $b = 20$ for the first public safety user, $a = 1$, $b = 30$ for the second public safety user. We set the application target rate $r_{i,s}^{t}$ for these two users to equal b that is 20 and 30, respectively. Another two normalized utility functions are used with the same a and b parameters to represent two commercial users running real-time applications. Each sigmoidal-like function is an approximation to a step function at rate b. We also use two logarithmic functions

Algorithm 7 Public safety UE algorithm

Send initial bid $w_i(1)$ to eNodeB
Send the application target rate $r_{i,s}^{t}$ to eNodeB
loop
  while Flag $\sum_{i=1}^{M} r_{i,s}^{t} \geq R$ from eNodeB **do**
    Receive shadow price $p(n)$ from eNodeB
    if STOP from eNodeB **then**
      Calculate allocated rate $r_{i,s}^{\text{opt}} = \frac{w_i(n)}{p(n)}$
    else
      Solve $r_{i,s}(n) = \arg\max_{r_{i,s}} \Big(\log U_i(r_{i,s}) - p(n) r_{i,s} \Big)$
      Send new bid $w_i(n) = p(n) r_{i,s}(n)$ to eNodeB
    end if
  end while
  while Flag $\sum_{i=1}^{M} r_{i,s}^{t} < R$ from eNodeB **do**
    Receive shadow price $p(n)$ from eNodeB
    if STOP from eNodeB **then**
      Calculate allocated rate $r_{i,s}^{\text{opt}} = \frac{w_i(n)}{p(n)}$
    else
      Solve $r_{i,s}(n) = r_i + r_{i,s}^{t} = \arg\max_{r_i} \Big(\log U_i(r_i + c_i) - p(n)(r_i + c_i) \Big)$
      Send new bid $w_i(n) = p(n)(r_i(n) + c_i)$ to eNodeB
    end if
  end while
end loop

Algorithm 8 Commercial UE algorithm

Send initial bid $w_i(1)$ to eNodeB
loop
  while Flag $\sum_{i=1}^{M} r_{i,s}^{t} \geq R$ from eNodeB **do**
    Allocated rate $r_{i,c}^{\text{opt}} = 0$
  end while
  while Flag $\sum_{i=1}^{M} r_{i,s}^{t} < R$ from eNodeB **do**
    Receive shadow price $p(n)$ from eNodeB
    if STOP from eNodeB **then**
      Calculate allocated rate $r_{i,c}^{\text{opt}} = \frac{w_i(n)}{p(n)}$
    else
      Solve $r_{i,c}(n) = \arg\max_{r_{i,c}} \Big(\log U_i(r_{i,c}) - p(n) r_{i,c} \Big)$
      Send new bid $w_i(n) = p(n) r_{i,c}(n)$ to eNodeB
    end if
  end while
end loop

Algorithm 9 eNodeB algorithm

```
loop
  Receive bids w_i(n) from UEs {Let w_i(0) = 0 ∀i}
  Receive application target rates from public safety UES
  while Σ_{i=1}^{M} r^t_{i,s} ≥ R do
    Send flag Σ_{i=1}^{M} r^t_{i,s} ≥ R to all UEs
    if |w_i(n) − w_i(n − 1)| < δ, i = {1, ...., M} then
      STOP and allocate rates (i.e. r^opt_{i,s} to public safety user i)
    else
      Calculate p(n) = (Σ_{i=1}^{M} w_i(n)) / R, i = {1, ...., M}
      Send new shadow price p(n) to public safety UEs
    end if
  end while
  while Σ_{i=1}^{M} r^t_{i,s} < R do
    Send flag Σ_{i=1}^{M} r^t_{i,s} < R to all UEs
    if |w_i(n) − w_i(n − 1)| < δ ∀i then
      STOP and allocate rates (i.e. r^opt_{i,s} or r^opt_{i,c} to user i)
    else
      Calculate p(n) = (Σ_{i=1}^{M+N} w_i(n)) / R
      Send new shadow price p(n) to all UEs
    end if
  end while
end loop
```

expressed in Eq. (2.2) with different parameters $k = 3$ for one public safety UE and $k = 0.5$ for second public safety UE running delay-tolerant application. We set the application target rate $r^t_{i,s}$ for each of these two users to equal 15. Another two logarithmic utility functions are used with the same k parameters to represent two commercial users running delay-tolerant applications.

4.1.4.1 Convergence Dynamics When $\sum_{i=1}^{M} r^t_{i,s} \geq R$

This represents the first case where $\sum_{i=1}^{M} r^t_{i,s} \geq R$. We set $R = 70$ and $\delta = 10^{-2}$. As mentioned before, in this case the commercial UEs will not be allocated any of the eNodeB resources because R does not exceed the public safety application target rates which need to be satisfied before the eNodeB starts allocating resources to the commercial users. In Fig. 4.1, we show the simulation results for the rate of different public safety users and the number of iterations. The sigmoidal-like utility functions are given priority over the logarithmic utility functions for rate allocation. This explains the results we got in Fig. 4.1. In this case the final optimal rate does not exceed the user application target rate. In Fig. 4.2, we show the bids of the four public safety users with the number of iterations. As expected, user rates are proportional to the user bids. The algorithm allows users with real-time applications to bid higher than the other users until each one of them reaches its inflection point, which is equivalent to their application target rates, then users with elastic traffic

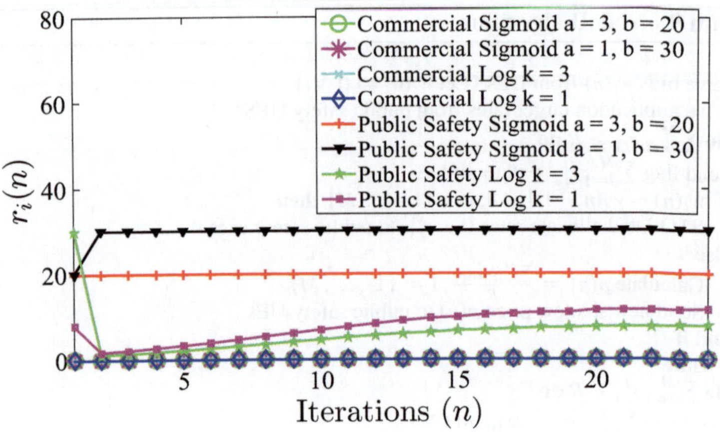

Fig. 4.1 The rates $r_i(n)$ with the number of iterations n for different users and $R = 70$

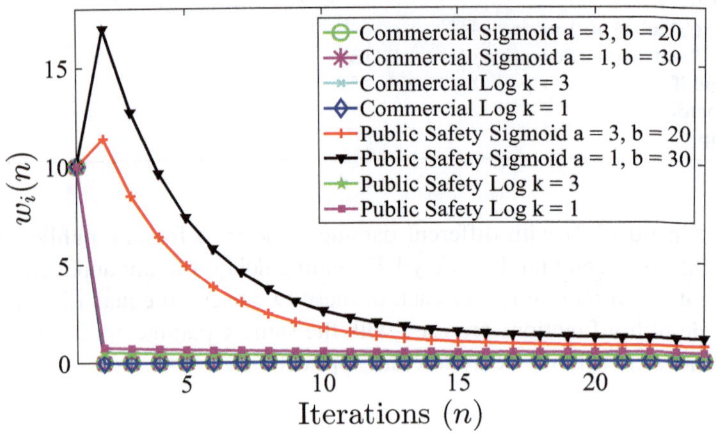

Fig. 4.2 The bids convergence $w_i(n)$ with the number of iterations n for different users and $R = 70$

start dividing the remaining resources among them based on their parameters while not exceeding their application target rates. In Fig. 4.3, we show the shadow price $p(n)$ with the number of iterations where the convergence behavior of the shadow price with the number of iterations is shown.

4.1.4.2 Convergence Dynamics When $\sum_{i=1}^{M} r_{i,s}^{t} < R$

Figure 4.4 shows four public safety normalized sigmoidal-like utility functions expressed in Eq. (2.1) corresponding to two public safety users and another two commercial users. We also show four logarithmic functions expressed in Eq. (2.2),

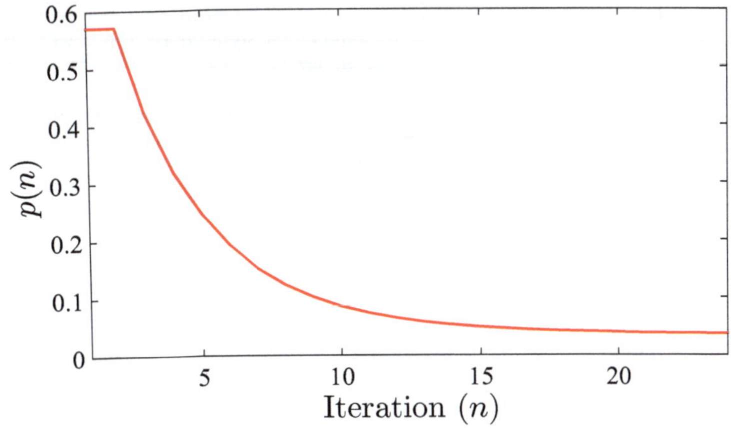

Fig. 4.3 The shadow price convergence with the number of iterations n

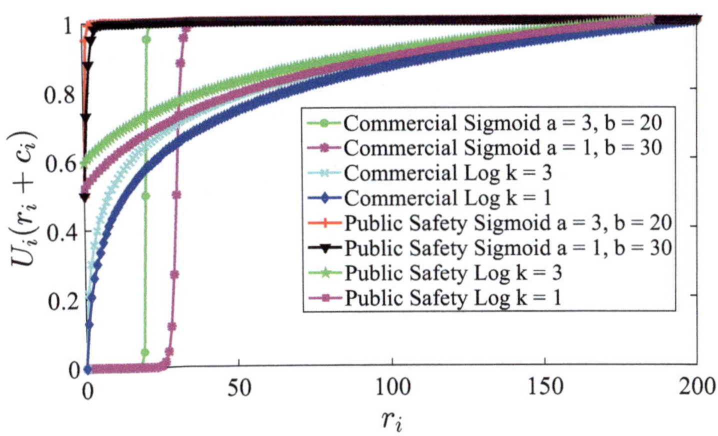

Fig. 4.4 The users utility functions $U_i(r_i + c_i)$

which represent delay-tolerant applications for two public safety users and another two commercial users. We set $R = 120$ and $\delta = 10^{-2}$. This represents the second case where $\sum_{i=1}^{M} r_{i,s}^{t} < R$. In this case the public safety UEs are given priority over the commercial UEs. In Fig. 4.5, we show the simulation results for the rate of different public safety and commercial users and the number of iterations. First the algorithm allocates an equivalent amount of resources to the application target rate to each public safety user. It then starts allocating resources to each commercial UE with inelastic traffic until it reaches the inflection point of that user utility function. It then starts dividing the remaining resources among all users based on their parameters. In Fig. 4.6, we show the bids of the eight users with the number of iterations. The algorithm allows public safety users to bid higher than the other users

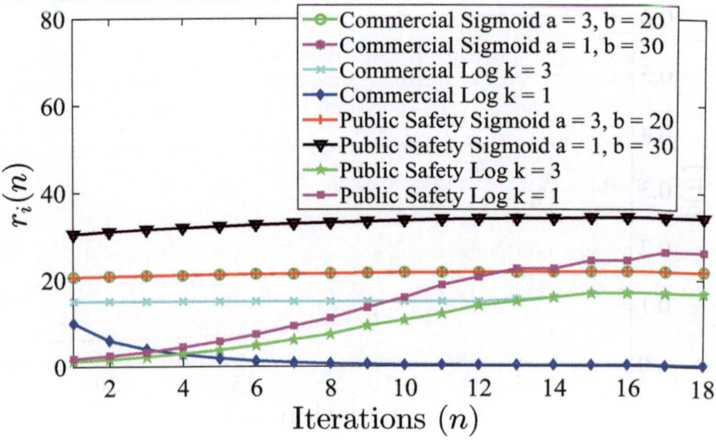

Fig. 4.5 The rates $r_i(n)$ with the number of iterations n for different users and $R = 200$

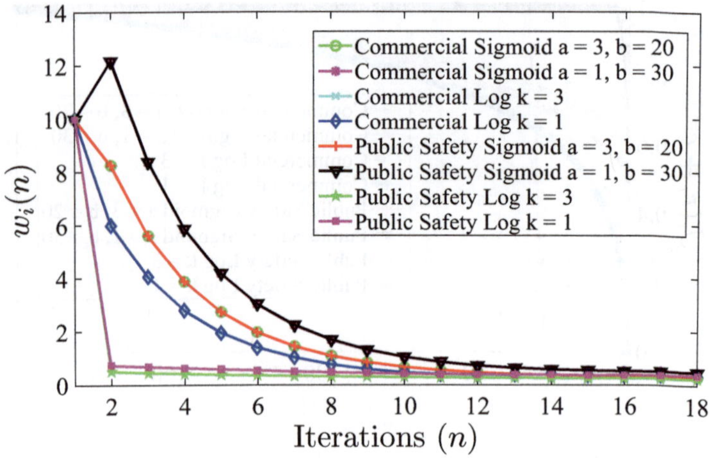

Fig. 4.6 The bids convergence $w_i(n)$ with the number of iterations n for different users and $R = 200$

until each one of them reaches its application target rate. Each commercial user with inelastic traffic then starts bidding higher until its application utility function reaches its inflection point. In Fig. 4.7, we show the shadow price $p(n)$ with the number of iterations where the convergence behavior of the shadow price with the number of iterations is shown.

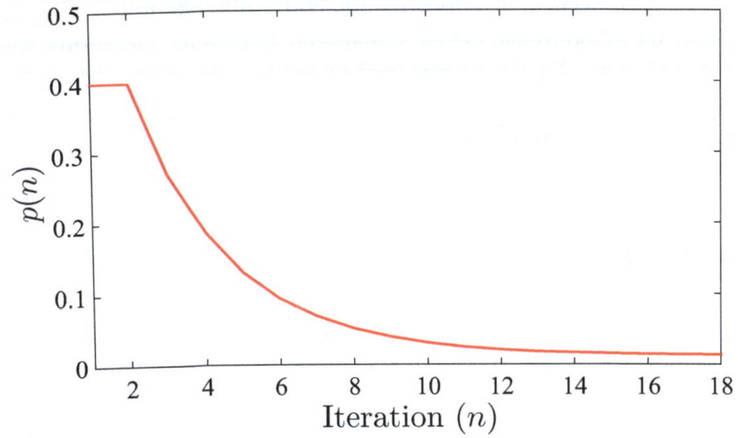

Fig. 4.7 The shadow price convergence with the number of iterations n

4.2 Multi-Application Resource Allocation with User Discrimination in Cellular Networks

In this section, we focus on finding an optimal solution for the resource allocation problem for different types of users running multiple types of applications simultaneously on their UEs. We considered subscriber differentiation, application status differentiation (application weight), and application target rate when formulating the resource allocation optimization problem. In our model, each user subscribing for a service is assigned a subscription weight by the network. Each user can run multiple applications simultaneously and each application is represented by a utility function based on the application type. In addition, each application is assigned an application weight by the UE based on the application instantaneous usage percentage and importance to the UE. Furthermore, certain type of users with higher priority (e.g., VIP users) are assigned applications target rates by the network. Therefore, these VIP UEs' applications are given higher priority by the network when allocating resources. A minimum QoS is guaranteed for each user by using a proportional fairness approach and real-time applications are given priority over delay-tolerant applications. Our objective is to allocate the resources optimally among the UEs and their applications from a single eNodeB based on a utility proportional fairness policy. We propose a two-stage rate allocation algorithm to allocate the eNodeB resources among users and their applications. In the first stage, the eNodeB collaborates with the UEs to allocate user rates. In the second stage, the rates are allocated to user applications internally by the UEs. The contributions in this section are summarized as:

- We present a resource allocation optimization problem to allocate the eNodeB resources optimally among different types of users running multiple applications.

- We propose a two-stage rate allocation method to allocate rates optimally among users. First, the eNodeB and the UE collaborate to allocate an optimal rate to each UE. Each UE then allocates its assigned rate optimally among its applications.
- We show that our resource allocation optimization problems have unique tractable global optimal solutions.

4.2.1 Problem Formulation

We consider a single cell mobile system that consists of a single eNodeB, M regular UEs, and another N VIP UEs as shown in Fig. 4.8. The rate allocated by the eNodeB to the ith UE is given by r_i. Each UE has its own utility function $X_i(r_i)$ that corresponds to the user satisfaction with its allocated rate r_i. Our objective is first to determine the optimal rates the eNodeB shall allocate to the UEs. We assume that the utility function $X_i(r_i)$ that is assigned to the ith user is given by:

$$X_i(r_i) = \prod_{j=1}^{L_i} U_{ij}^{\alpha_{ij}}(r_{ij} + c_{ij}) \tag{4.5}$$

$$c_{ij} = \begin{cases} r_{ij}^t & \text{if the } j\text{th application is assigned} \\ & \text{an application target rate} \\ 0 & \text{if the } j\text{th application is not assigned} \\ & \text{an application target rate,} \end{cases}$$

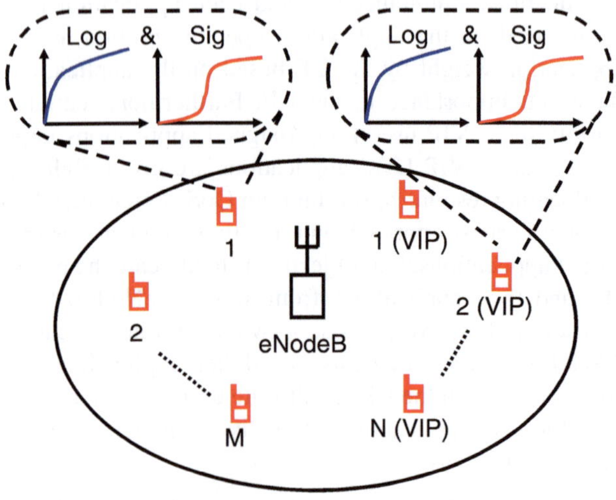

Fig. 4.8 System model, one eNodeB with N VIP UEs and another M regular UEs subscribing for a mobile service in the eNodeB coverage area

where $U_{ij}(r_{ij})$ is the jth application utility function for user i, r_{ij} is the rate allocated to the jth application running on the ith UE, L_i is the number of applications running on the ith UE, c_{ij} is the application target rate for the jth application of user i if it is assigned one whereas it is 0 if the jth application is not assigned an application target rate by the network, α_{ij} is the jth application usage percentage (application weight) of the ith UE and r^t_{ij} is the application target rate assigned to the jth application of the ith user.

We express the user satisfaction with its provided service using utility functions [13, 21]. We assume that the jth application utility function for user i is given by $U_{ij}(r_{ij})$ that is strictly concave function expressed by Eq. (2.2) or sigmoidal-like function expressed by Eq. (2.1) where r_{ij} is the rate allocated to the jth application of user i. Delay-tolerant applications are represented by logarithmic utility functions whereas real-time applications are represented by sigmoidal-like utility functions.

4.2.2 *Resource Allocation Optimization Problem*

The resource allocation optimization problem for multi-application users is divided into two cases. The first case is when the maximum available resources R of the eNodeB is less than or equal to the total VIP UEs applications target rates. The second case is when R is greater than the total UEs applications target rates. The RA optimization problems for the two cases will be solved by our proposed algorithm to obtain the optimal rate for each UE as well as the optimal rates for the UE applications.

4.2.2.1 First-Case RA Optimization Problem When $\sum_{i=1}^{M} \sum_{j=1}^{L_i} r^t_{ij} \geq R$

In this case, the eNodeB only allocates resources to the M VIP UEs as they are considered more important and regular users will not be allocated any of the eNodeB resources since its available resources are limited. In this case, the optimization problem is divided into two stages. In the first stage, the eNodeB allocates rates r_i to the M group of users. Both the eNodeB and the M UEs collaborate to achieve the UEs resource allocation. In the second stage, each one of these M UEs uses the rate allocated to it by the eNodeB to allocate optimal rates r_{ij} to its L_i applications. The second stage is performed internally in the UE.

First Stage of the First-Case Optimization Problem

In this case, the optimization problem for the first stage can be written as:

$$\max_{\mathbf{r}} \quad \prod_{i=1}^{M} X_i^{\beta_i}(r_i)$$
$$\text{subject to} \quad \sum_{i=1}^{M} r_i \le R \tag{4.6}$$
$$0 \le r_i \le \sum_{j=1}^{L_i} r_{ij}^{t}, \quad i = 1, 2, \ldots, M$$

where $X_i = \prod_{j=1}^{L_i} U_{ij}^{\alpha_{ij}}(r_{ij})$, $\mathbf{r} = \{r_1, r_2, \ldots, r_M\}$ is the rate allocated by the eNodeB to the ith UE, M is the number of VIP UEs in the coverage area of the eNodeB, R is the maximum achievable rate of the given eNodeB, and β_i is the ith user subscription weight assigned by the network.

The objective function in the optimization problem (4.6) is equivalent to $\sum_{i=1}^{M} \beta_i \log(X_i(r_i))$. Therefore, the optimization problem (4.6) is a convex optimization problem and there exists a unique tractable global optimal solution as shown in Corollary (III.1) [31]. This optimal solution gives each of the M users an optimal rate r_i^{opt} that is less than or equal to the total applications target rates for that UE.

Second Stage of the First-Case Optimization Problem

Each one of the M VIP UEs allocates optimal rates r_{ij}^{opt} to its L_i applications. The optimal rate allocated to each application depends on the application differentiation weight and the application type. This optimization problem is solved internally in the UE and can be written for the ith UE as follows:

$$\max_{\mathbf{r}_i} \quad \prod_{j=1}^{L_i} U_{ij}^{\alpha_{ij}}(r_{ij})$$
$$\text{subject to} \quad \sum_{j=1}^{L_i} r_{ij} \le r_i^{\text{opt}} \tag{4.7}$$
$$0 \le r_{ij} \le r_{ij}^{\text{t}}, \quad j = 1, 2, \ldots, L_i$$

where $\mathbf{r}_i = \{r_{i1}, r_{i2}, \ldots, r_{iL_i}\}$, r_i^{opt} is the optimal rate allocated by the eNodeB to the ith UE and L_i is the number of UE applications. Since the objective function in the optimization problem (4.7) is equivalent to $\sum_{j=1}^{L_i} \alpha_{ij} \log(U_{ij}(r_{ij}))$, then optimization problem (4.7) is convex and there exists a unique tractable global optimal solution as shown in Corollary (III.2) [31]. This optimal solution represents the optimal rate r_{ij}^{opt} allocated to each of the L_i applications.

4.2.2.2 Second-Case RA Optimization Problem When $\sum_{i=1}^{M}\sum_{j=1}^{L_i} r_{ij}^t < R$

In this case, the eNodeB first allocates resources to the M VIP UEs. It then allocates the remaining resources based on the proportional fairness approach. The optimization problem in this case is divided into two stages. In the first stage, the eNodeB collaborates with the UEs to allocate rates r_i to all UEs. In the second stage, each one of these $M + N$ UEs allocates optimal rates r_{ij} to its applications. The second stage is performed internally in the UE. The inelastic traffic is given priority when allocating the resources internally by the UEs.

First Stage of the Second-Case Optimization Problem

In this case, the optimization problem of the first stage can be written as:

$$\begin{aligned} \max_{\mathbf{r}} \quad & \prod_{i=1}^{M+N} X_i^{\beta_i}(r_i) \\ \text{subject to} \quad & \sum_{i=1}^{M+N} r_i \le R \\ & r_i \ge 0, \quad i = 1, 2, \ldots, M+N \end{aligned} \tag{4.8}$$

where $X_i = \prod_{j=1}^{L_i} U_{ij}^{\alpha_{ij}}(r_{ij}+c_{ij})$, $\mathbf{r} = \{r_1, r_2, \ldots, r_{M+N}\}$, $M+N$ is the number of VIP and regular UEs subscribing for a service in the coverage area of the eNodeB and β_i is the ith user subscription weight assigned by the network. Each UE is allocated at least the total amount of its applications target rates if it has any.

The objective function in the optimization problem (4.8) is equivalent to $\sum_{i=1}^{M+N} \beta_i \log(X_i(r_i))$. Therefore, optimization problem (4.8) is a convex optimization problem and there exists a unique tractable global optimal solution r_i^{opt} for each of the $M + N$ users as shown in Corollary (III.1) [31].

Second-Stage of the Second-Case Optimization Problem

Each one of the $M + N$ UEs allocates optimal rates r_{ij}^{opt} to its applications. Each UE first allocates the application target rate to each of its applications if it is assigned one. It then starts allocating the remaining resources among all the applications based on the application differentiation weight and the type of the application. This optimization problem is solved internally in the UE and can be written for the ith UE as follows:

$$\begin{aligned} \max_{\mathbf{r}_i} \quad & \prod_{j=1}^{L_i} U_{ij}^{\alpha_{ij}}(r_{ij} + c_{ij}) \\ \text{subject to} \quad & \sum_{j=1}^{L_i} (r_{ij} + c_{ij}) \leq r_i^{\text{opt}} \\ & r_{ij} \geq 0, \quad j = 1, 2, \ldots, L_i \end{aligned} \tag{4.9}$$

where $\mathbf{r}_i = \{r_{i1}, r_{i2}, \ldots, r_{iL_i}\}$, r_i^{opt} is the rate allocated by the eNodeB to the ith UE in the first stage and c_{ij} is same as before. The objective function of the optimization problem (4.9) is equivalent to $\sum_{j=1}^{L_i} \alpha_{ij} \log(U_{ij}(r_{ij} + c_{ij}))$. Therefore, optimization problem (4.9) is a convex optimization problem and there exists a unique tractable global optimal solution as shown in Corollary (III.2) [31]. Each UE allocates an optimal rate $r_{ij}^{\text{opt}} = r_{ij} + c_{ij}$ to each of its applications.

4.2.3 *Resource Allocation with User Discrimination Algorithms*

As mentioned before, the RA for the multi-application users with different priorities is achieved in two stages. In the first stage, the eNodeB and the UEs collaborate to allocate optimal rates r_i for users as shown in VIP UE Algorithm 10, regular UE Algorithm 11, and eNodeB Algorithm 12. In the second stage, the UE internal algorithm allocates applications rates r_{ij} to the UE's applications as shown in the internal UE algorithm 13.

4.2.3.1 First-Stage RA Algorithm

The first stage of the RA algorithm is presented in this section. The algorithm starts when each UE transmits an initial bid $w_i(1)$ to the eNodeB. Additionally, each VIP UE transmits its applications target rates to the eNodeB. The eNodeB checks whether the $\sum_{i=1}^{M} \sum_{i=1}^{L_i} r_{ij}^t$ is less or greater than R and sends a flag with this information to each UE. In the case of $\sum_{i=1}^{M} \sum_{i=1}^{L_i} r_{ij}^t \geq R$, the regular UEs will not be allocated any of the resources and will not be sending any further bids to the eNodeB.

Each VIP UE checks whether the difference between the current received bid and the previous one is less than a threshold δ, if so it exits. Otherwise, the eNodeB calculates the shadow price $p(n) = \frac{\sum_{i=1}^{M} w_i(n)}{R}$ and sends it to the VIP UEs where it is used to calculate the ith VIP UE rate $r_i(n)$ which is the solution of the optimization problem $r_i(n) = \arg\max_{r_i}(\beta_i \log X_i(r_i) - p(n)r_i)$ where $X_i(r_i) = \prod_{j=1}^{L_i} U_{ij}^{\alpha_{ij}}(r_{ij})$. A new bid $w_i(n) = p(n)r_i(n)$ is then calculated and the VIP UEs check the fluctuation condition as in [15] and send their new bids to the eNodeB. The algorithm is finalized by the eNodeB. Each VIP UE then calculates its allocated rate $r_i^{\text{opt}} = \frac{w_i(n)}{p(n)}$.

Algorithm 10 VIP UE algorithm

Send initial bid $w_i(1)$ to eNodeB
Send the applications target rates r_{ij}^t to eNodeB
loop
  while Flag $\sum_{i=1}^{M}\sum_{j=1}^{L_i} r_{ij}^t \geq R$ from eNodeB **do**
    Receive shadow price $p(n)$ from eNodeB
    if STOP from eNodeB **then**
      Calculate allocated rate $r_i^{\text{opt}} = \frac{w_i(n)}{p(n)}$
    else
      Solve $r_i(n) = \arg\max_{r_i}\left(\beta_i \log X_i(r_i) - p(n) r_i\right)$
      Send new bid $w_i(n) = p(n) r_i(n)$ to eNodeB
    end if
  end while
  while Flag $\sum_{i=1}^{M}\sum_{j=1}^{L_i} r_{ij}^t < R$ from eNodeB **do**
    Receive shadow price $p(n)$ from eNodeB
    if STOP from eNodeB **then**
      Calculate allocated rate $r_i^{\text{opt}} = \frac{w_i(n)}{p(n)}$
    else
      Solve $r_i(n) = \arg\max_{r_i}\left(\beta_i \log X_i(r_i) - p(n)(r_i + \sum_{j=1}^{L_i} r_{ij}^t)\right)$
      Calculate new bid $w_i(n) = p(n)(r_i(n) + \sum_{j=1}^{L_i} r_{ij}^t)$
      if $|w_i(n) - w_i(n-1)| > \Delta w$ **then**
        $w_i(n) = w_i(n-1) + \text{sign}(w_i(n) - w_i(n-1))\Delta w(n)$
        $\{\Delta w(n) = l_1 e^{-\frac{n}{l_2}}\}$
      end if
      Send new bid $w_i(n)$ to eNodeB
    end if
  end while
end loop

In the case of $\sum_{i=1}^{M}\sum_{i=1}^{L_i} r_i^t < R$, a flag with this information is sent to each UE by the eNodeB. Each UE checks whether the difference between the current received bid and the previous one is less than a threshold δ, if so it exits. Otherwise, the eNodeB calculates the shadow price $p(n) = \frac{\sum_{i=1}^{M+N} w_i(n)}{R}$ and sends it to each UE where it is used by the VIP UE to calculate the rate $r_i = r_i(n) + \sum_{j=1}^{L_i} r_{ij}^{\text{t}}$, $r_i(n)$ which is the solution of the optimization problem $r_i(n) = \arg\max_{r_i}(\beta_i \log X_i(r_i) - p(n)(r_i + \sum_{j=1}^{L_i} r_{ij}^{\text{t}}))$ where $X_i(r_i) = \prod_{j=1}^{L_i} U_{ij}^{\alpha_{ij}}(r_{ij} + c_{ij})$. A new bid $w_i(n) = p(n)(r_i(n) + \sum_{j=1}^{L_i} r_{ij}^{\text{t}})$ is calculated by the VIP UE. All VIP UEs check the fluctuation condition and send their new bids to the eNodeB. On the other hand, the regular UEs receive $p(n)$ and calculate the rate $r_i(n)$ which is the solution of the optimization problem $r_i(n) = \arg\max_{r_i}(\beta_i \log X_i(r_i) - p(n) r_i)$ where $X_i(r_i) = \prod_{j=1}^{L_i} U_{ij}^{\alpha_{ij}}(r_{ij} + c_{ij})$. A new bid $w_i(n) = p(n) r_i(n)$ is calculated by the regular UE. All regular UEs check the fluctuation condition and send their new bids to the eNodeB. The algorithm is finalized by the eNodeB. Each VIP and regular UE then calculates its allocated rate $r_i^{\text{opt}} = \frac{w_i(n)}{p(n)}$.

Algorithm 11 Regular UE algorithm

Send initial bid $w_i(1)$ to eNodeB
loop
  while Flag $\sum_{i=1}^{M}\sum_{j=1}^{L_i} r_{ij}^t \geq R$ from eNodeB **do**
    Allocated rate $r_i^{\text{opt}} = 0$
  end while
  while Flag $\sum_{i=1}^{M}\sum_{j=1}^{L_i} r_{ij}^t < R$ from eNodeB **do**
    Receive shadow price $p(n)$ from eNodeB
    if STOP from eNodeB **then**
      Calculate allocated rate $r_i^{\text{opt}} = \frac{w_i(n)}{p(n)}$
    else
      Solve $r_i(n) = \arg\max_{r_i}\left(\beta_i \log X_i(r_i) - p(n)r_i\right)$
      Calculate new bid $w_i(n) = p(n)r_i(n)$
      if $|w_i(n) - w_i(n-1)| > \Delta w$ **then**
        $w_i(n) = w_i(n-1) + \text{sign}(w_i(n) - w_i(n-1))\Delta w(n)$
        $\{\Delta w(n) = l_1 e^{-\frac{n}{l_2}}\}$
      end if
      Send new bid $w_i(n)$ to eNodeB
    end if
  end while
end loop

Algorithm 12 eNodeB algorithm

---

loop
  Receive bids $w_i(n)$ from UEs $\{\text{Let } w_i(0) = 0 \;\forall i\}$
  Receive applications target rates from VIP UEs
  while $\sum_{i=1}^{M}\sum_{j=1}^{L_i} r_{ij}^t \geq R$ **do**
    Send flag $\sum_{i=1}^{M}\sum_{j=1}^{L_i} r_{ij}^t \geq R$ to all UEs
    if $|w_i(n) - w_i(n-1)| < \delta,\ i = \{1,\ldots,M\}$ **then**
      STOP and allocate rates (i.e. r_i^{opt} to VIP user i)
    else
      Calculate $p(n) = \frac{\sum_{i=1}^{M} w_i(n)}{R},\ i = \{1,\ldots,M\}$
      Send new shadow price $p(n)$ to VIP UEs
    end if
  end while
  while $\sum_{i=1}^{M}\sum_{j=1}^{L_i} r_{ij}^t < R$ **do**
    Send flag $\sum_{i=1}^{M}\sum_{j=1}^{L_i} r_{ij}^t < R$ to all UEs
    if $|w_i(n) - w_i(n-1)| < \delta \;\forall i$ **then**
      STOP and allocate rates (i.e. r_i^{opt} to user i)
    else
      Calculate $p(n) = \frac{\sum_{i=1}^{M+N} w_i(n)}{R}$
      Send new shadow price $p(n)$ to all UEs
    end if
  end while
end loop

Algorithm 13 Internal UE algorithm

loop
  Receive r_i^{opt} from eNodeB Algorithm 10, 11 and 12
  Solve
  $\mathbf{r}_i = \arg\max_{\mathbf{r}_i} \sum_{j=1}^{L_i} (\alpha_{ij} \log U_{ij}(r_{ij} + c_{ij}) - p(r_{ij} + c_{ij})) + pr_i^{\text{opt}}$
  $\{\mathbf{r}_i = \{r_{i1}, r_{i2}, \ldots, r_{iL_i}\}\}$
  Allocate $r_{ij}^{\text{opt}} = r_{ij} + c_{ij}$ to the jth application
end loop

4.2.3.2 Second-Stage RA Algorithm

The second stage of RA is presented in this section and shown in Algorithm 13 where the rates r_{ij} are allocated internally by the UE to its applications. Each UE uses its allocated rate r_i^{opt} in the first stage to solve the optimization problem $\mathbf{r}_i = \arg\max_{\mathbf{r}_i} \sum_{j=1}^{L_i} (\alpha_{ij} \log U_{ij}(r_{ij} + c_{ij}) - p(r_{ij} + c_{ij})) + pr_i^{\text{opt}}$. The rate $r_{ij}^{\text{opt}} = r_{ij} + c_{ij}$ is then allocated to the UE's jth application.

4.2.4 Numerical Results

In this section, we consider one eNodeB with four UEs in its coverage area subscribing for a mobile service. The first and second UEs are VIP UEs and the third and fourth UEs are regular UEs. Each one of the four UEs is running two applications simultaneously. The first application is a real-time application whereas the second application is a delay-tolerant application.

We applied Algorithms 10–13 in C++ (Sects. 4.5.2 and 4.5.3) to the UEs functions. The simulation results showed convergence to the optimal global point in the two stages of the algorithm. We present the simulation results for the four users. The first UE is a VIP UE, we use a normalized sigmoidal-like utility function that is expressed by Eq. (2.1) to represent its first application with $a = 3$, $b = 20$ which is an approximation to a step function at rate $r = 20$ and we set $r_{11}^t = 20$. Additionally, for the second application of the first user (VIP user) we use a logarithmic function that is expressed by Eq. (2.2) with $k = 3$ which is an approximation of a delay-tolerant application. The second user is a VIP user, we use a normalized sigmoidal-like utility function to represent its first application with $a = 1, b = 30$ and we set $r_{21}^t = 30$. Additionally, for the second application of the second user (VIP user) we use a logarithmic function with $k = 0.5$ to represent its delay-tolerant application. The same parameters of the first user are used for the third user's utility functions except that its applications are not assigned applications target rates. Also, the same parameters of the second user are used for the fourth user's utility functions except that its applications are not assigned applications target rates. Furthermore, we set $\beta_i = 1$ for all UEs. We use $r_{\max} = 100$ for all

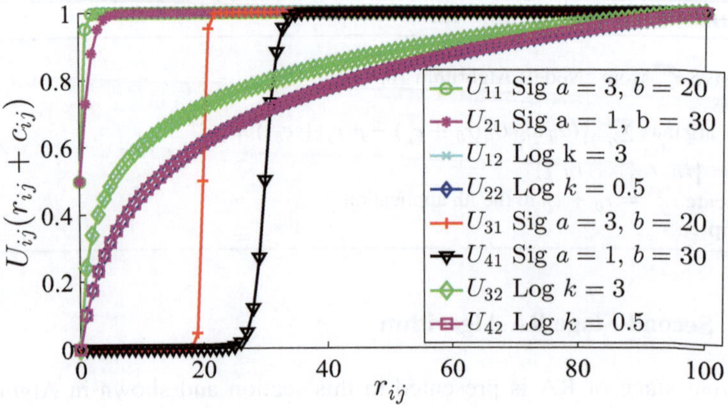

Fig. 4.9 The applications utility functions $U_{ij}(r_{ij})$

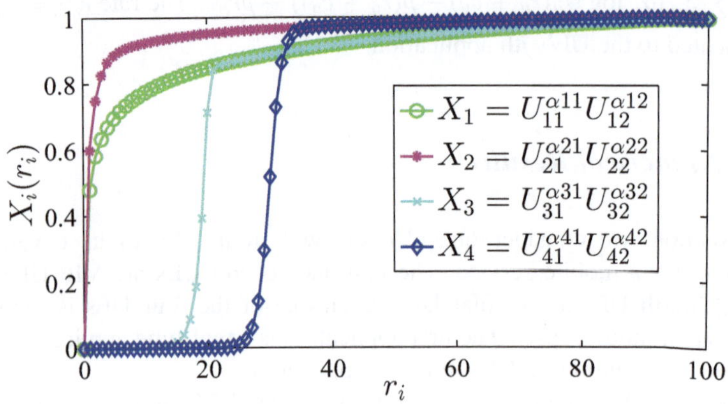

Fig. 4.10 The aggregated utility functions $X_i(r_i)$ of the ith user

logarithmic functions, $l_1 = 5$ and $l_2 = 10$ in the fluctuation decay function of the algorithm and $\delta = 10^{-3}$. Let the application weight α_{ij} in the set α corresponds to the jth application of user i where $\alpha = \{\alpha_{11}, \alpha_{12}, \alpha_{21}, \alpha_{22}, \alpha_{31}, \alpha_{32}, \alpha_{41}, \alpha_{42}\}$.

Figure 4.9 shows eight applications utility functions corresponding to the four UEs. The real-time applications of the VIP UEs are assigned applications target rates, this explains their shifted utility functions by the amount of r_{ij}^t in Fig. 4.9. The other applications do not have applications target rates ($c_{ij} = 0$ for each one). Figure 4.10 shows the aggregated utilities $X_i(r_i)$ for each user.

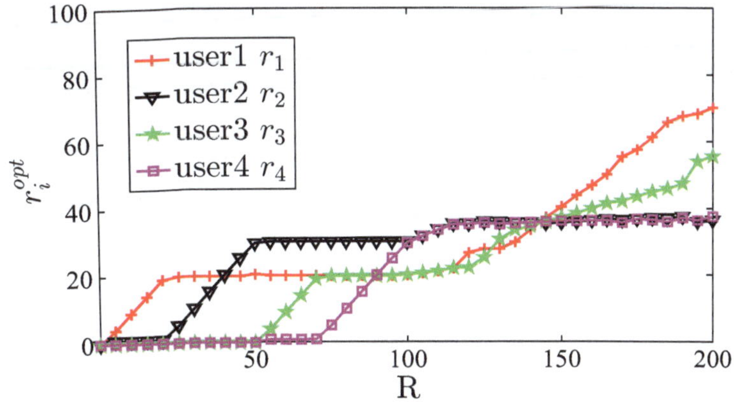

Fig. 4.11 The users optimal rates r_i^{opt} for different values of R

4.2.4.1 Convergence Dynamics for Different Values of *R*

In the following simulations, we set $\alpha = \{0.5, 0.5, 0.9, 0.1, 0.5, 0.5, 0.9, 0.1\}$ and the eNodeB available resources R takes values between 5 and 200 with step of 5. In Fig. 4.11, we show the four users optimal rates r_i^{opt} with different eNodeB resources R. This represents the solution of optimization problem (4.6) when $R \leq 50$ and optimization problem (4.8) when $R > 50$, using the first stage of the algorithm, where 50 is the total applications target rates for the two VIP users. Figure 4.11 shows that when $R \leq 50$ the regular UEs are not allocated any of the eNodeB resources. Furthermore, when $R > 50$ each VIP user is first allocated its total applications target rates and the remaining resources are then allocated to all users based on the proportional fairness approach.

In Fig. 4.12, we show the final optimal applications rates r_{ij}^{opt} for the four users with different eNodeB resources R. This is the solution of optimization problem (4.7) when $R \leq 50$ and the solution of (4.9) when $R > 50$ using the user internal algorithm. The figure shows that when $R \leq 50$, the real-time applications are given priority over the delay-tolerant applications when allocating rates by each VIP UE to its applications whereas when $R > 50$, the VIP UEs first allocate the applications target rates to the applications that are assigned ones and then allocate the remaining resources among all applications using proportional fairness approach while giving the priority to the real-time applications. The regular users also give the priority to their real-time applications when allocating resources as shown in the same figure.

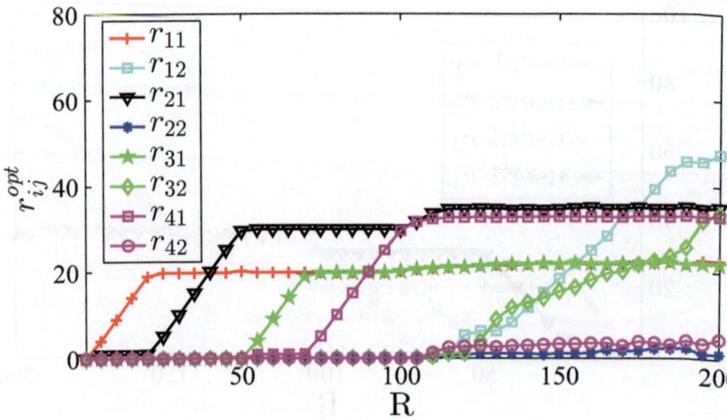

Fig. 4.12 The applications optimal rates r_{ij}^{opt} for different values of R

4.2.4.2 Rate Allocation Sensitivity to Change in α

In the following simulations, we measure the sensitivity of the change in application weight that is corresponding to the application usage percentage in the UE. We use $R = 200$ and the same parameters as before for the four users. The users change their applications usage percentage with time as follows:

$$\alpha(t) = \begin{cases} \alpha = \{0.1, 0.9, 0.5, 0.5, 0.9, 0.1, 0.5, 0.5\}; \\ \hfill \mathit{for}\ \ 0 \le t \le 10 \\ \alpha = \{0.5, 0.5, 0.3, 0.7, 0.2, 0.8, 0.1, 0.9\}; \\ \hfill \mathit{for}\ \ 10 \le t \le 20 \\ \alpha = \{1.0, 0.0, 0.9, 0.1, 0.8, 0.2, 0.1, 0.9\}; \\ \hfill \mathit{for}\ \ 20 \le t \le 30 \end{cases} . \tag{4.10}$$

Figure 4.13 shows the users optimal rates r_i^{opt} with time for the changing usage percentages given by $\alpha(t)$.

4.2.5 *Price Sensitivity to Change in R*

In the following simulations, we set $\alpha = \{0.5, 0.5, 0.9, 0.1, 0.5, 0.5, 0.9, 0.1\}$ and the eNodeB available resources takes different values between 150 and 350 with step of 20. In Fig. 4.14, we show the shadow price p, that represents the total price per unit bandwidth for all users, with eNodeB available resources R when the number of users is fixed. As expected the price is higher for smaller R when the number of users is fixed. This is equivalent to high price for high traffic and low price for low traffic.

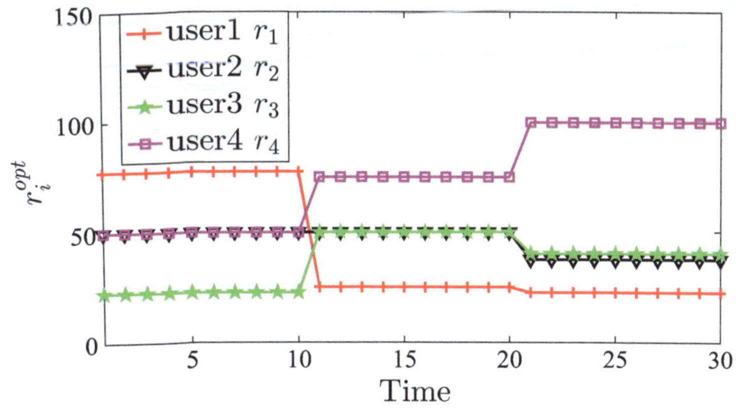

Fig. 4.13 The users optimal rates r_i^{opt} with the change in users' applications usage percentages $\alpha(t)$

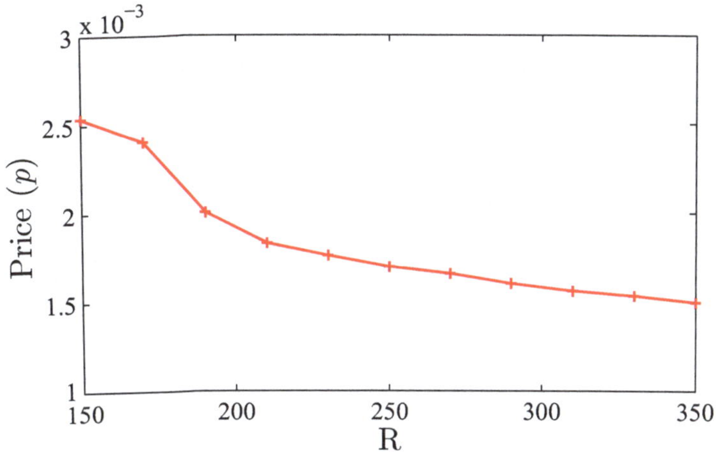

Fig. 4.14 The shadow price p for different values of R and fixed number of users

4.2.6 Price Sensitivity to Change in the Number of UEs

In the following simulations, we set $\alpha = \{0.5, 0.5, 0.9, 0.1, 0.5, 0.5, 0.9, 0.1\}$ for each group of users where each group is the same four users as before. The number of users takes values in multiples of four between 4 and 20 where each time we run the algorithm additional 4 users, with the same original users applications' parameters, are added. We use $R = 2000$ for the eNodeB available resources. In Fig. 4.15, we show the shadow price p, with the number of users. When the number of users increases (high traffic) the price increases (i.e., R is fixed). It is obvious that users are encouraged to run their non-urgent applications during the low traffic periods for less price.

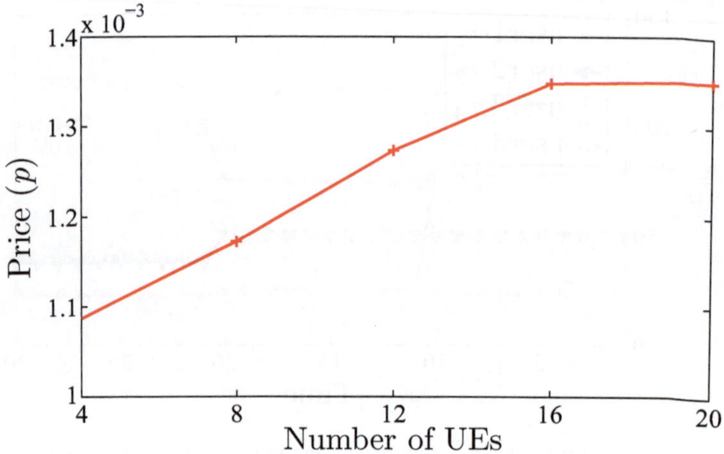

Fig. 4.15 The shadow price p for different number of users and $R = 2000$

4.3 Resource Allocation with User Discrimination for Multi-Carrier Cellular Networks

In this section, we provide an efficient framework for the resource allocation problem to allocate multi-carrier resources optimally among users that belong to different classes of user groups. In our model, we use utility functions to represent users' applications. Sigmoidal-like utility functions and logarithmic utility functions are used to represent real-time and delay-tolerant applications, respectively, running on the UEs [14]. The resource allocation with user discrimination framework presented in [2] does not consider the case of multi-carrier resources available at the eNodeB. It only solves the problem of resource allocation with user discrimination in the case of single carrier. In this section, we consider the case of multiple carriers' resources available at the eNodeB and multiple classes of users located under the coverage area of these carriers. We use a priority criterion for the resource allocation process that varies based on the user's class and the type of application running on the UE. We consider two classes of users, VIP users (i.e., public safety users or users who require emergency services) and regular users. VIP users are assigned a minimum required application rate for each of their applications whereas regular users' applications are not assigned any.

We formulate the resource allocation with user discrimination problem in a multi-stage resource allocation with carrier aggregation optimization problem to allocate resources to each user from its all in range carriers based on a utility proportional fairness policy. Each application running on the UE is assigned an application minimum required rate by the network that varies based on the type of user's application and the user's class. Furthermore, if the user's in range carriers have enough available resources, the user is allocated at minimum its applications' minimum required rates. VIP users are given priority over regular users by the network when allocating each carrier's resources, and real-time applications are given priority over delay-tolerant applications. The contributions in this section are summarized as:

- We present a multi-stage resource allocation with user discrimination optimization problem to allocate multi-carrier resources optimally among different classes of users.
- We prove that the resource allocation optimization problem is convex and therefore the global optimal solution is tractable.
- We present a resource allocation algorithm to solve the optimization problem and allocate each user an aggregated final rate from its in range carriers. The proposed algorithm outperforms that presented in [2] as it considers allocating each user resources from multiple carriers using a resource allocation with carrier aggregation approach.
- We present simulation results for the performance of the proposed resource allocation algorithm.

4.3.1 Multi-Carrier Problem

In this paper, we consider a single cell mobile system with one eNodeB, K carriers (frequency bands) that have resources available at the eNodeB, M regular, and VIP UEs. Let $\mathscr{M}$ be the set of all regular and VIP UEs where $M = |\mathscr{M}|$. The set of carriers is given by $\mathscr{K} = \{1, 2, \ldots, K\}$ with carriers in order from the highest frequency to the lowest frequency. Higher frequency carriers have smaller coverage area than lower frequency carriers. The eNodeB allocates resources from multiple carriers to each UE. Users located under the coverage area of multiple carriers are allocated resources from all in range carriers. The rate allocated by the eNodeB to UE i from all in range carriers is given by r_i. Each application running on the UE is mathematically represented by a utility function $U_i(r_i)$ that corresponds to the application's type and represents the user satisfaction with its allocated rate r_i. Our goal is to determine the optimal rates that the eNodeB shall allocate from each carrier to each UE in order to maximize the total system utility while ensuring proportional fairness between utilities.

The rate allocated to the ith user in $\mathscr{M}$ by the jth carrier in $\mathscr{K}$ is given by $r_i^{j,\text{all}}$. The final allocated rate by the eNodeB to the ith user is given by

$$r_i = \sum_{j \in \mathscr{K}} r_i^{j,\text{all}} \tag{4.11}$$

where r_i is equivalent to the sum of rates allocated to the ith user from all carriers in its range. Based on the coverage area of each carrier and the users' classes, a user grouping method is introduced in Sect. 4.3.1.1 to partition users into groups.

The eNodeB performs resource allocation with user discrimination based on carrier aggregation to allocate each carrier's resources to users located within the coverage area of that carrier.

We express the user satisfaction with its rate using utility functions that represent the degree of satisfaction of the user function with the rate allocated by the cellular network [13, 15, 21]. We represent the ith user application utility function $U_i(r_i)$ by sigmoidal-like function expressed by Eq. (2.1) or logarithmic function expressed by Eq. (2.2) where r_i is the rate of the ith user.

4.3.1.1 User Grouping Method

In this section we introduce a user grouping method to create user groups for each carrier $j \in \mathcal{K}$. The eNodeB creates a user group $\mathcal{M}_j$ for each carrier where $\mathcal{M}_j$ is a set of users located under the coverage area of the jth carrier. The number of users in $\mathcal{M}_j$ is given by $M_j = |\mathcal{M}_j|$. Furthermore, users in $\mathcal{M}_j$ are partitioned into two groups of users. A VIP user group $\mathcal{M}_j^{\text{VIP}}$ and a regular user group $\mathcal{M}_j^{\text{Reg}}$, where $\mathcal{M}_j^{\text{VIP}}$ and $\mathcal{M}_j^{\text{Reg}}$ are the sets of all VIP users and regular users, respectively, located under the coverage area of the jth carrier with $\mathcal{M}_j = \mathcal{M}_j^{\text{VIP}} \cup \mathcal{M}_j^{\text{Reg}}$. The number of users in $\mathcal{M}_j^{\text{VIP}}$ and $\mathcal{M}_j^{\text{Reg}}$ is given by $M_j^{\text{VIP}} = |\mathcal{M}_j^{\text{VIP}}|$ and $M_j^{\text{Reg}} = |\mathcal{M}_j^{\text{Reg}}|$, respectively. The eNodeB allocates the jth carrier resources to users in $\mathcal{M}_j$ with a priority given to VIP users (i.e., users in $\mathcal{M}_j^{\text{VIP}}$). Users located under the coverage area of multiple carriers (i.e., common users in multiple user groups) are allocated resources from these carriers and their final rates are aggregated under a non-adjacent inter-band aggregation scenario.

The ith user is considered part of user group $\mathcal{M}_j$ if it is located within a distance of D_j from the eNodeB where D_j represents the coverage radius of the jth carrier. Let d_i denotes the distance between the eNodeB and user i. The jth carrier user group $\mathcal{M}_j$ is defined as

$$\mathcal{M}_j = \{i : d_i < D_j, 1 \leq i \leq M\}, \quad 1 \leq j \leq K. \tag{4.12}$$

On the other hand, the eNodeB creates a set of carriers $\mathcal{K}_i$, for each user, that is defined as

$$\mathcal{K}_i = \{j : d_i < D_j, 1 \leq j \leq K\}, \quad 1 \leq i \leq M. \tag{4.13}$$

The number of carriers that the ith user can be allocated resources from is given by $N_i = |\mathcal{K}_i|$. Higher frequency carriers have smaller coverage radius than lower frequency carriers (i.e., $D_1 < D_2 < \cdots < D_K$). Therefore, user group $\mathcal{M}_1 \subseteq \mathcal{M}_2 \subseteq \cdots \subseteq \mathcal{M}_K$. Figure 4.16 shows one cellular cell with one eNodeB under non-adjacent inter-band scenario with K carriers in $\mathcal{K}$ and M users in $\mathcal{M}$ and how users are partitioned into user groups based on their location and their class.

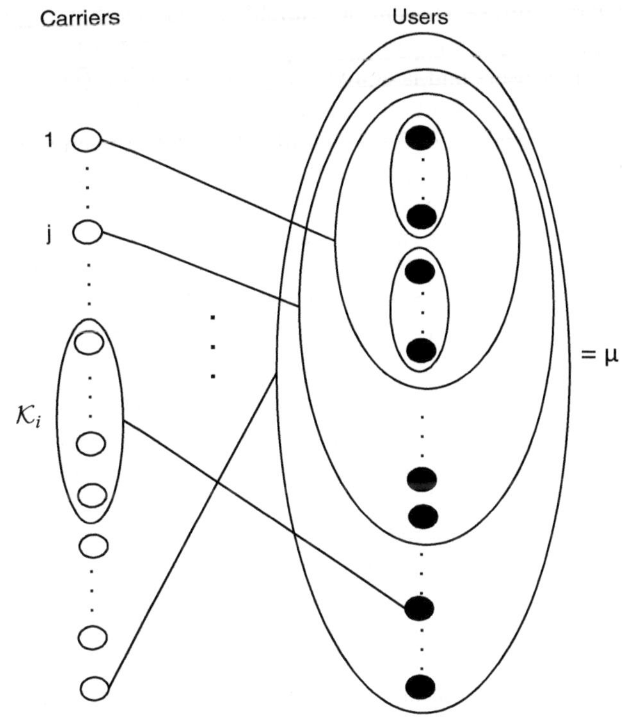

Fig. 4.16 User grouping for an LTE mobile system with M users in $\mathcal{M}$ and K carriers in $\mathcal{K}$. $\mathcal{M}_j$ represents the set of users located under the coverage area of the jth carrier with $\mathcal{M}_j = \mathcal{M}_j^{\text{VIP}} \cup \mathcal{M}_j^{\text{Reg}}$. $\mathcal{K}_i$ represents the set of all in range carriers for the ith user

4.3.2 *Multi-Carrier RA with User discrimination Optimization Problem*

In this section, we present a multi-stage resource allocation (RA) with user discrimination optimization problem to allocate multi-carrier resources optimally among users in their coverage area. Our objective is to find the final allocated rate to each user from its all in range carriers based on a utility proportional fairness policy. We use utility functions of users rates to represent the type of application running on the UE. Every user subscribing for a mobile service is guaranteed to achieve a minimum QoS with priority criterion. VIP users are given priority when allocating each carrier's resources and within each user class group, whether it is VIP or regular user group, real-time applications are given priority when allocating each carrier's resources. This is due to the nature of sigmoidal-like utility functions that are used to represent real-time applications.

The eNodeB performs the resource allocation process for all carriers one at a time and one after another in ascending order of their coverage radius D_j. Each carrier $j \in \mathscr{K}$ has a limited amount of available resources that is given by R_j and each user's application has a minimum required rate r_i^{req} that is equivalent to zero in the case of regular users and is equivalent to certain value (i.e., rate) in the case of VIP users. The eNodeB starts the RA process by performing a RA for carrier 1 in $\mathscr{K}$ as it has the smallest coverage radius D_1. After allocating its resources to users in $\mathscr{M}_1$, the eNodeB then starts the RA process to allocate carrier 2 resources to users in $\mathscr{M}_2$. In addition, since $\mathscr{M}_1 \subseteq \mathscr{M}_2$ the eNodeB allocates users in $\mathscr{M}_1$ resources from carrier 2 and the rates are aggregated based on a non-adjacent inter-band aggregation scenario. The eNodeB continues the resource allocation process by allocating the jth carrier resources to users in $\mathscr{M}_j$. Let $r_i^{j,\text{all}}$ represents the rate allocated by the jth carrier to UE i and let C_i represents the total aggregated rate allocated to UE i by carriers $\{1,2,\ldots,j-1\}$ where $C_i = \sum_{l=1}^{j-1} r_i^{l,\text{all}}$. Furthermore, let C_i^j be a constant that is always equivalent to zero for regular users whereas for VIP users C_i^j is equivalent to zero or $r_i^{\text{req}} - C_i$ based on some conditions that are discussed later in this section. The resource allocation process is finalized by allocating the Kth carrier resources to users in $\mathscr{M}_K$, i.e., all users in the cellular cell as they are all located within its coverage radius. We consider a utility proportional fairness objective function, based on carrier aggregation, that the eNodeB seeks to maximize for each time it allocates the carrier's resources.

The proposed RA optimization problem for multi-carrier cellular systems is divided into three cases. In order for the eNodeB to guarantee that VIP users are given priority when allocating each carrier's resources, each time the eNodeB performs a RA process for a carrier it checks the values of (1) the carrier's available resources R_j, (2) the current total rate allocated to each VIP UE $i \in \mathscr{M}_j^{\text{VIP}}$ from other carriers (i.e., $C_i = \sum_{l=1}^{j-1} r_i^{l,\text{all}}$), and (3) the value of $r_i^{\text{req}} - C_i$ for each VIP UE $i \in \mathscr{M}_j^{\text{VIP}}$ if $C_i < r_i^{\text{req}}$. Based on these values, the eNodB performs the RA process that corresponds to the most appropriate case among the three cases. The three cases and their RA optimization framework are presented below.

Case 1. RA Optimization Problem When $C_i \geq r_i^{\text{req}}$ $\forall i \in \mathscr{M}_j$

The eNodeB chooses the RA optimization problem of this case in order to allocate the jth carrier resources if the total aggregated rate C_i that is allocated to each UE $i \in \mathscr{M}_j$ from carriers $\{1,2,\ldots,j-1\}$ is greater than or equal to the minimum required application rate r_i^{req}. In this case, since each UE has already been allocated at least its application minimum required rate from other carriers, the eNodeB performs the RA process among all users under the coverage area of carrier j. The RA optimization problem for the jth carrier in this case is given by:

$$\begin{aligned}
\max_{\mathbf{r}^j} \quad & \prod_{i=1}^{M_j} U_i(C_i + C_i^j + r_i^j) \\
\text{subject to} \quad & \sum_{i=1}^{M_j} r_i^{j,\text{all}} \le R_j, \quad r_i^{j,\text{all}} \ge 0 \\
& r_i^{j,\text{all}} = r_i^j + C_i^j, \quad C_i^j = 0 \\
& C_i = \sum_{l=1}^{j-1} r_i^{l,\text{all}}, \quad C_i \ge r_i^{\text{req}}, \quad i = 1, 2, \ldots, M_j,
\end{aligned} \tag{4.14}$$

where C_i^j is a constant that is equivalent to zero in this case, $U_i(C_i + C_i^j + r_i^j)$ is the utility function of the summation of the rate C_i allocated to the application running on the ith user by carriers $\{1, 2, \ldots, j-1\}$ and the rate $r_i^{j,\text{all}}$ allocated to the same application by carrier j where $r_i^{j,\text{all}} = C_i^j + r_i^j$, $\mathbf{r}^j = \{r_1^j, r_2^j, \ldots, r_{M_j}^j\}$ and M_j is the number of users in $\mathcal{M}_j$ (i.e., both VIP and regular users) located under the coverage area of the jth carrier. After the eNodeB performs the RA process for the jth carrier by solving optimization problem (4.14), the total rate allocated to each user by the eNodeB is equivalent to $C_i + r_i^{j,\text{all}}$. In optimization problem (4.14), we consider a utility proportional fairness objective function, based on carrier aggregation, that the eNodeB seeks to maximize when it performs RA for carrier j.

Case 2. RA Optimization Problem When $C_i < r_i^{\text{req}}$ for Any User $i \in \mathcal{M}_j$ and $\sum_{i=1}^{M_j^{\text{VIP}}} q_i^j \ge R_j$ Where $q_i^j = 0$ if $C_i \ge r_i^{\text{req}}$ and $q_i^j = r_i^{\text{req}} - C_i$ if $C_i < r_i^{\text{req}}$
The eNodeB selects the optimization problem of this case to allocate the jth carrier resources if the total aggregated rate C_i for any user i is less than the user's application minimum required rate r_i^{req} and $\sum_{i=1}^{M_j^{\text{VIP}}} q_i^j$ for VIP users in $\mathcal{M}_j^{\text{VIP}}$ is greater than or equal to the carrier's available resources R_j. In this case, the eNodeB allocates the jth carrier resources only to VIP UEs in $\mathcal{M}_j^{\text{VIP}}$ as they are considered more important and regular users in $\mathcal{M}_j^{\text{Reg}}$ are not allocated any of the jth carrier resources since the carrier's resources are limited. The RA optimization problem for the jth carrier in this case is given by:

$$
\begin{aligned}
\max_{\mathbf{r}^j} \quad & \prod_{i=1}^{M_j^{\mathrm{VIP}}} U_i(C_i + c_i^j + r_i^j) \\
\text{subject to} \quad & \sum_{i=1}^{M_j^{\mathrm{VIP}}} r_i^{j,\mathrm{all}} \le R_j, \quad r_i^{j,\mathrm{all}} \ge 0 \\
& C_i = \sum_{l=1}^{j-1} r_i^{l,\mathrm{all}}, \quad r_i^{j,\mathrm{all}} = r_i^j + c_i^j \\
& c_i^j = 0 \\
& q_i^j = \begin{cases} 0 & \text{if } C_i \ge r_i^{\mathrm{req}} \\ r_i^{\mathrm{req}} - C_i & \text{if } C_i < r_i^{\mathrm{req}} \end{cases} \\
& \sum_{i=1}^{M_j^{\mathrm{VIP}}} q_i^j \ge R_j, \quad i = 1, 2, \ldots, M_j^{\mathrm{VIP}},
\end{aligned}
\tag{4.15}
$$

where $\mathbf{r}^j = \{r_1^j, r_2^j, \ldots, r_{M_j^{\mathrm{VIP}}}^j\}$, $c_i^j = 0$ and M_j^{VIP} is the number of users in $\mathcal{M}_j^{\mathrm{VIP}}$. After the eNodeB performs the RA process for the jth carrier by solving optimization problem (4.15), each VIP user in $\mathcal{M}_j^{\mathrm{VIP}}$ is allocated a rate that is equivalent to $r_i^{j,\mathrm{all}}$ by carrier j whereas users in $\mathcal{M}_j^{\mathrm{Reg}}$ are not allocated any of the jth carrier resources. The total rate allocated by the eNodeB to each user is equivalent to $C_i + r_i^{j,\mathrm{all}}$. In optimization problem (4.15), we consider a utility proportional fairness objective function, based on carrier aggregation, that the eNodeB seeks to maximize when it performs RA for carrier j.

Case 3. RA Optimization Problem When $C_i < r_i^{\mathrm{req}}$ for Any User $i \in \mathcal{M}_j^{\mathrm{VIP}}$ and $\sum_{i=1}^{M_j^{\mathrm{VIP}}} q_i^j < R_j$ Where $q_i^j = 0$ if $C_i \ge r_i^{\mathrm{req}}$ and $q_i^j = r_i^{\mathrm{req}} - C_i$ if $C_i < r_i^{\mathrm{req}}$

The eNodeB selects the optimization problem of this case to allocate the jth carrier resources if the total aggregated rate C_i for any user i is less than the user's application minimum required rate r_i^{req} and the summation $\sum_{i=1}^{M_j^{\mathrm{VIP}}} q_i^j$ for VIP users in $\mathcal{M}_j^{\mathrm{VIP}}$ is less than the carrier's available resources R_j. In this case, the eNodeB allocates the jth carrier resources to all UEs in $\mathcal{M}_j$. The RA optimization problem for the jth carrier in this case is given by:

$$
\begin{aligned}
\max_{\mathbf{r}^j} \quad & \prod_{i=1}^{M_j} U_i(C_i + C_i^j + r_i^j) \\
\text{subject to} \quad & \sum_{i=1}^{M_j} r_i^{j,\mathrm{all}} \leq R_j, \quad r_i^{j,\mathrm{all}} \geq 0 \\
& C_i = \sum_{l=1}^{j-1} r_i^{l,\mathrm{all}}, \quad r_i^{j,\mathrm{all}} = r_i^j + C_i^j \\
& C_i^j = \begin{cases} 0 & \text{if } C_i \geq r_i^{\mathrm{req}} \\ r_i^{\mathrm{req}} - C_i & \text{if } C_i < r_i^{\mathrm{req}} \end{cases} \\
& q_i^j = \begin{cases} 0 & \text{if } C_i \geq r_i^{\mathrm{req}} \\ r_i^{\mathrm{req}} - C_i & \text{if } C_i < r_i^{\mathrm{req}} \end{cases} \\
& \sum_{i=1}^{M_j^{\mathrm{VIP}}} q_i^j < R_j, \quad i = 1, 2, \ldots, M_j,
\end{aligned}
\tag{4.16}
$$

where $\mathbf{r}^j = \{r_1^j, r_2^j, \ldots, r_{M_j}^j\}$ and M_j is the number of users in $\mathscr{M}_j$. After the eNodeB performs the RA process for the jth carrier by solving optimization problem (4.16), each user in $\mathscr{M}_j$ is allocated a rate that is equivalent to $r_i^{j,\mathrm{all}}$ by carrier j and the total rate allocated by the eNodeB to each user is equivalent to $C_i + r_i^{j,\mathrm{all}}$. In optimization problem (4.16), we consider a utility proportional fairness objective function, based on carrier aggregation, that the eNodeB seeks to maximize when it performs RA for carrier j.

Each of the three RA optimization problems (4.14)–(4.16) of the jth carrier can be expressed by the following generalized optimization problem:

$$
\begin{aligned}
\max_{\mathbf{r}^j} \quad & \prod_{i=1}^{|\alpha_j|} U_i(C_i + C_i^j + r_i^j) \\
\text{subject to} \quad & \sum_{i=1}^{|\alpha_j|} r_i^{j,\mathrm{all}} \leq R_j, \quad r_i^{j,\mathrm{all}} \geq 0 \\
& C_i = \sum_{l=1}^{j-1} r_i^{l,\mathrm{all}}, \quad r_i^{j,\mathrm{all}} = r_i^j + C_i^j \\
& q_i^j = \begin{cases} 0 & \text{if } C_i \geq r_i^{\mathrm{req}} \\ r_i^{\mathrm{req}} - C_i & \text{if } C_i < r_i^{\mathrm{req}} \end{cases} \\
& i = 1, 2, \ldots, |\alpha_j|,
\end{aligned}
\tag{4.17}
$$

where C_i^j and α_j in (4.17) are given by

$$C_i^j = \begin{cases} 0 & \text{if } C_i \geq r_i^{\text{req}} \\ r_i^{\text{req}} - C_i & \text{if } C_i < r_i^{\text{req}} \text{ and } \sum_{i=1}^{|\mathscr{M}_j^{\text{VIP}}|} q_i^j < R_j \\ 0 & \text{if } C_i < r_i^{\text{req}} \text{ and } \sum_{i=1}^{|\mathscr{M}_j^{\text{VIP}}|} q_i^j \geq R_j \end{cases}$$

$$\alpha_j = \begin{cases} \mathscr{M}_j & \text{if } C_i \geq r_i^{\text{req}} \ \forall i \in \mathscr{M}_j \\ \mathscr{M}_j^{\text{VIP}} & \text{if } C_i < r_i^{\text{req}} \text{ for any user } i \in \mathscr{M}_j \\ & \text{and } \sum_{i=1}^{M_j^{\text{VIP}}} q_i^j \geq R_j \\ \mathscr{M}_j & \text{if } C_i < r_i^{\text{req}} \text{ for any user } i \in \mathscr{M}_j^{\text{VIP}} \\ & \text{and } \sum_{i=1}^{M_j^{\text{VIP}}} q_i^j < R_j \end{cases} \tag{4.18}$$

where $\mathbf{r}^j = \{r_1^j, r_2^j, \ldots, r_{|\alpha_j|}^j\}$, α_j is a set of users located under the coverage area of carrier j that is equivalent to $\mathscr{M}_j$ or $\mathscr{M}_j^{\text{VIP}}$ based on certain conditions as shown in (4.18) and $|\alpha_j|$ is the number of users in α_j.

The objective function in optimization problem (4.17) is equivalent to $\sum_{i=1}^{|\alpha_j|} \log U_i(C_i + C_i^j + r_i^j)$. Later in this section we prove that optimization problem (4.17) is a convex optimization problem and there exists a unique tractable global optimal solution. Once the eNodeB is done performing the RA process, for the jth carrier, by solving optimization problem (4.17), each user in α_j is allocated a rate that is equivalent to $r_i^{j,\text{all}} = r_i^j + C_i^j$ and the user's total aggregated rate allocated by the eNodeB from carriers $\{1, 2, \ldots, j\}$ is given by $\sum_{l=1}^{j} r_i^{l,\text{all}}$.

Lemma 4.3.1 *The utility functions* $\log U_i(C_i + C_i^j + r_i^j)$ *in optimization problem (4.17) are strictly concave functions.*

Proof The utility functions are assumed to be logarithmic functions expressed by Eq. (2.2) or sigmoidal-like functions expressed by Eq. (2.1). Therefore, $U_i(C_i + C_i^j + r_i^j)$ is a strictly concave (i.e., in the case of logarithmic utility functions) or a sigmoidal-like function of the total aggregated rate $C_i + C_i^j + r_i^j$ allocated to user i application from carriers $\{1, 2, \ldots, j\}$ after performing the RA process of the jth carrier by the eNodeB.

In the case of logarithmic utility function, recall the utility function properties in Sect. 2.1 of Chap. 2, the utility function of the application rate is positive, increasing and twice differentiable with respect to the application rate. It follows that $U_i'(C_i + C_i^j + r_i^j) = \frac{dU_i(C_i + C_i^j + r_i^j)}{dr_i^j} > 0$ and $U_i''(C_i + C_i^j + r_i^j) = \frac{d^2 U_i(C_i + C_i^j + r_i^j)}{dr_i^{j^2}} < 0$, i.e., since $C_i + C_i^j$ is greater or equal to zero. Then the function $\log U_i(C_i +$

$C_i^j + r_i^j)$ has $\frac{d\log(U_i(C_i+C_i^j+r_i^j))}{dr_i^j} = \frac{U_i'(C_i+C_i^j+r_i^j)}{U_i(C_i+C_i^j+r_i^j)} > 0$ and $\frac{d^2\log(U_i(C_i+C_i^j+r_i^j))}{dr_i^{j2}} = \frac{U_i''(C_i+C_i^j+r_i^j)U_i(C_i+C_i^j+r_i^j)-U_i'^2(C_i+C_i^j+r_i^j)}{U_i^2(C_i+C_i^j+r_i^j)} < 0$. Therefore, the natural logarithm of the logarithmic utility function $\log(U_i(C_i + C_i^j + r_i^j))$ is strictly concave.

On the other hand, in the case of sigmoidal-like utility function, the normalized sigmoidal-like function is given by $U_i(C_i + C_i^j + r_i^j) = c_i\Big(\frac{1}{1+e^{-a_i(C_i+C_i^j+r_i^j-b_i)}} - d_i\Big)$. For $0 < r_i^j < (R_j - C_i^j)$, we have

$$0 < c_i\Big(\frac{1}{1+e^{-a_i(C_i+C_i^j+r_i^j-b_i)}} - d_i\Big) < 1$$

$$d_i < \frac{1}{1+e^{-a_i(C_i+C_i^j+r_i^j-b_i)}} < \frac{1+c_id_i}{c_i}$$

$$\frac{1}{d_i} > 1+e^{-a_i(C_i+C_i^j+r_i^j-b_i)} > \frac{c_i}{1+c_id_i}$$

$$0 < 1-d_i(1+e^{-a_i(C_i+C_i^j+r_i^j-b_i)}) < \frac{1}{1+c_id_i}$$

It follows that for $0 < r_i^j < (R_j - C_i^j)$, we have the first and second derivatives as

$$\frac{d}{dr_i^j}\log U_i(C_i + C_i^j + r_i^j) = \frac{a_id_ie^{-a_i(C_i+C_i^j+r_i^j-b_i)}}{1-d_i(1+e^{-a_i(C_i+C_i^j+r_i^j-b_i)})} + \frac{a_ie^{-a_i(C_i+C_i^j+r_i^j-b_i)}}{(1+e^{-a_i(C_i+C_i^j+r_i^j-b_i)})} > 0$$

$$\frac{d^2}{dr_i^{j2}}\log U_i(C_i + C_i^j + r_i^j) = \frac{-a_i^2d_ie^{-a_i(C_i+C_i^j+r_i^j-b_i)}}{c_i\Big(1-d_i(1+e^{-a(C_i+C_i^j+r_i^j-b_i)})\Big)^2} + \frac{-a_i^2e^{-a_i(C_i+C_i^j+r_i^j-b_i)}}{(1+e^{-a_i(C_i+C_i^j+r_i^j-b_i)})^2} < 0.$$

Therefore, the natural logarithm of the sigmoidal-like utility function $\log(U_i(C_i + C_i^j + r_i^j))$ is strictly concave function. Therefore, the utility functions natural logarithms have strictly concave natural logarithms in both cases of logarithmic utility functions and sigmoidal-like utility functions.

Theorem 4.3.2 proves the convexity of optimization problem (4.17).

Theorem 4.3.2 *Optimization problem (4.17) is a convex optimization problem and there exists a unique tractable global optimal solution.*

Proof It follows from Lemma 4.3.1 that all UEs utility functions of applications rates are strictly concave. Therefore, optimization problem (4.17) is a convex optimization problem. For a convex optimization problem there exists a unique tractable global optimal solution [28–30].

4.3.3 Multi-Carrier Optimization Algorithm

In this section, we present our multi-carrier resource allocation with user discrimination algorithm. The proposed algorithm consists of UE and eNodeB parts shown in Algorithms 14 and 15, respectively. The execution of the algorithm starts by UEs, subscribing for mobile services, transmitting their application utility parameters to the eNodeB, which allocates available carriers' resources to UEs based on a proportional fairness policy. First, the eNodeB performs the user grouping method described in Sect. 4.3.1.1 for each carrier by creating three user group sets $\mathcal{M}_j^{\text{VIP}}$, $\mathcal{M}_j^{\text{Reg}}$, and $\mathcal{M}_j$ for UEs located within the coverage area of the jth carrier. It then starts performing the RA process to allocate the carriers resources starting with carrier 1 in $\mathcal{K}$ (i.e., the carrier with the smallest coverage radius) in ascending order $1 \rightarrow K$. In order to allocate certain carrier's resources, the eNodeB performs the RA process that corresponds to the most appropriate case among the three cases presented in Sect. 4.3.2. From optimization problem (4.17), we have the following Lagrangian:

$$L(\mathbf{r}^j, p^j) = \sum_{i=1}^{|\alpha_j|} \log U_i(C_i + C_i^j + r_i^j) - p^j \left(\sum_{i=1}^{|\alpha_j|} (C_i^j + r_i^j) + \sum_{i=1}^{|\alpha_j|} z_i - R_j \right), \tag{4.19}$$

where $z_i \geq 0$ is the slack variable and p^j is Lagrange multiplier that represents the shadow price (price per unit bandwidth for all the $|\alpha_j|$ channels). The rates, solutions to Eq. (4.17), are the values r_i^j which solve equation $\frac{\partial \log U_i(C_i+C_i^j+r_i^j)}{\partial r_i^j} = p^j$ and are the intersection of the time varying shadow price, horizontal line $y = p^j$, with the curve

Algorithm 14 The ith UE algorithm

loop
  Send application utility parameters k_i, a_i, b_i, $r_i^{\max}$ and r_i^{req} to eNodeB.
  Receive the final allocated rate r_i from the eNodeB.
end loop

$y = \frac{\partial \log U_i(C_i+C_i^j+r_i^j)}{\partial r_i^j}$ geometrically. The rate allocated by carrier j to the ith UE is equivalent to $r_i^{j,\text{all}} = r_i^j + C_i^j$. When the eNodeB is done allocating the Kth carrier resources, each user is then allocated its final aggregated rate $r_i = \sum_{j=1}^{K} r_i^{j,\text{all}}$.

Algorithm 15 The eNodeB algorithm

loop
  Initialize $C_i = 0$; $C_i^j = 0$; $r_i^{j,\text{all}} = 0$.
  Receive application utility parameters $k_i, a_i, b_i, r_i^{\max}$ and r_i^{req} from all UEs in $\mathcal{M}$.
  for $j \leftarrow 1$ to K **do**
    Create user groups $\mathcal{M}_j^{\text{VIP}}$, $\mathcal{M}_j^{\text{Reg}}$ and $\mathcal{M}_j$ for UEs located within the coverage area of the jth carrier.
  end for
  for $i \leftarrow 1$ to $|\mathcal{M}_j|$ **do**
    Create carrier group $\mathcal{K}_i$ for the ith UE's all in range carriers.
  end for
  for $j \leftarrow 1$ to K **do**
    if $C_i < r_i^{\text{req}}$ **then**
      $q_i^j = r_i^{\text{req}} - C_i$
    else
      $q_i^j = 0$
    end if
    if $C_i \geq r_i^{\text{req}}$ $\forall i \in \mathcal{M}_j$ **then**
      $C_i^j = 0$
      Solve $\mathbf{r}^j = \arg\max_{\mathbf{r}^j} \sum_{i=1}^{|\mathcal{M}_j|} \log U_i(C_i + C_i^j + r_i^j) - p^j(\sum_{i=1}^{|\mathcal{M}_j|}(r_i^j + C_i^j) - R_j)$.
      Allocate rate $r_i^{j,\text{all}} = r_i^j + C_i^j$ by the jth carrier to each user in $\mathcal{M}_j$.
      Calculate new $C_i = C_i + r_i^{j,\text{all}}$ $\forall i \in \mathcal{M}_j$
    else if $C_i < r_i^{\text{req}}$ for any user $i \in \mathcal{M}_j$ && $\sum_{i=1}^{M_j^{\text{VIP}}} q_i^j \geq R_j$ **then**
      $C_i^j = 0$
      Solve $\mathbf{r}^j = \arg\max_{\mathbf{r}^j} \sum_{i=1}^{|\mathcal{M}_j^{\text{VIP}}|} \log U_i(C_i + C_i^j + r_i^j) - p^j(\sum_{i=1}^{|\mathcal{M}_j^{\text{VIP}}|}(r_i^j + C_i^j) - R_j)$.
      Allocate rate $r_i^{j,\text{all}} = r_i^j + C_i^j$ by the jth carrier to each user in $\mathcal{M}_j^{\text{VIP}}$.
      Calculate new $C_i = C_i + r_i^{j,\text{all}}$ $\forall i \in \mathcal{M}_j^{\text{VIP}}$
    else if $C_i < r_i^{\text{req}}$ for any user $i \in \mathcal{M}_j^{\text{VIP}}$ and $\sum_{i=1}^{|\mathcal{M}_j^{\text{VIP}}|} q_i^j < R_j$ **then**
      if $C_i < r_i^{\text{req}}$ **then**
        $C_i^j = r_i^{\text{req}} - C_i$
      else
        $C_i^j = 0$
      end if
      Solve $\mathbf{r}^j = \arg\max_{\mathbf{r}^j} \sum_{i=1}^{|\mathcal{M}_j|} \log U_i(C_i + C_i^j + r_i^j) - p^j(\sum_{i=1}^{|\mathcal{M}_j|}(r_i^j + C_i^j) - R_j)$.
      Allocate rate $r_i^{j,\text{all}} = r_i^j + C_i^j$ by the jth carrier to each user in $\mathcal{M}_j$.
      Calculate new $C_i = C_i + r_i^{j,\text{all}}$ $\forall i \in \mathcal{M}_j$
    end if
  end for
  Allocate total aggregated rate $r_i = \sum_{j=1}^{K} r_i^{j,\text{all}}$ by the eNodeB to each UE i in $\mathcal{M}$
end loop

4.3.4 Numerical Results

Algorithms 14 and 15 were applied in C++ (Sect. 4.5.4) to multiple utility functions with different parameters. Simulation results showed convergence to the global optimal rates. In this section, we consider a mobile cell with one eNodeB, two carriers with available resources and eight active UEs located under the coverage area of the eNodeB as shown in Fig. 4.17. The UEs are divided into two groups. The first group of UEs (index $i = \{1, 2, 3, 4\}$) represents user group $\mathcal{M}_1$ located within the coverage radius D_1 of carrier 1. Each user in $\mathcal{M}_1$ belongs to one of the two classes of user groups, i.e., VIP user group and regular user group, where $\mathcal{M}_1^{\text{VIP}} = \{2, 4\}$, $\mathcal{M}_1^{\text{Reg}} = \{1, 3\}$, and $\mathcal{M}_1 = \mathcal{M}_1^{\text{VIP}} \cup \mathcal{M}_1^{\text{Reg}}$. On the other hand, the second group of UEs (index $i = \{1, 2, 3, 4, 5, 6, 7, 8\}$) represents user group $\mathcal{M}_2$ located within the coverage radius D_2 of carrier 2. Each user in $\mathcal{M}_2$ belongs to a VIP user group or a regular user group where $\mathcal{M}_2^{\text{VIP}} = \{2, 4, 6, 8\}$, $\mathcal{M}_2^{\text{Reg}} = \{1, 3, 5, 7\}$, and $\mathcal{M}_2 = \mathcal{M}_2^{\text{VIP}} \cup \mathcal{M}_2^{\text{Reg}}$.

We use sigmoidal-like utility functions and logarithmic utility functions with different parameters to represent each of the users' applications. We use three normalized sigmoidal-like functions that are expressed by Eq. (2.1) with different parameters. The used parameters are $a_i = 5$, $b_i = 10$ that correspond to a sigmoidal-like function with inflection point $r_i = 10$ which represents the utility of UE with index $i = \{5\}$, $a_i = 3$, $b_i = 20$ that correspond to a sigmoidal-like function with inflection point $r_i = 20$ which represents the utility of UE with index $i = \{1\}$, and $a_i = 1$, $b_i = 30$ that correspond to a sigmoidal-like function with inflection point $r_i = 30$ which represents the utility of UEs with indexes $i = \{2, 6\}$, as shown in Fig. 4.18. We use three logarithmic functions expressed by Eq. (2.2) with $r_i^{\max} = 100$ and different k_i parameters to represent delay-tolerant applications. We

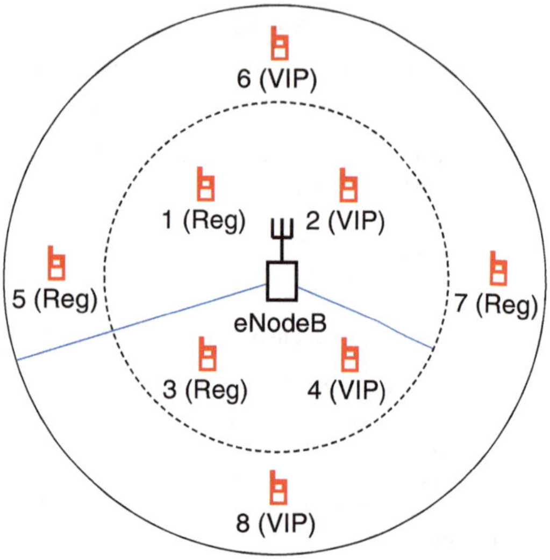

Fig. 4.17 System model for a mobile system with $M = 8$ users and $K = 2$ carriers available at the eNodeB. Carrier 1 coverage radius is D_1 and carrier 2 coverage radius is D_2 with $D_1 < D_2$. $\mathcal{M}_1 = \{1, 2, 3, 4\}$ and $\mathcal{M}_2 = \{1, 2, \ldots, 8\}$ represent the sets of user groups located under the coverage area of carrier 1 and carrier 2, respectively

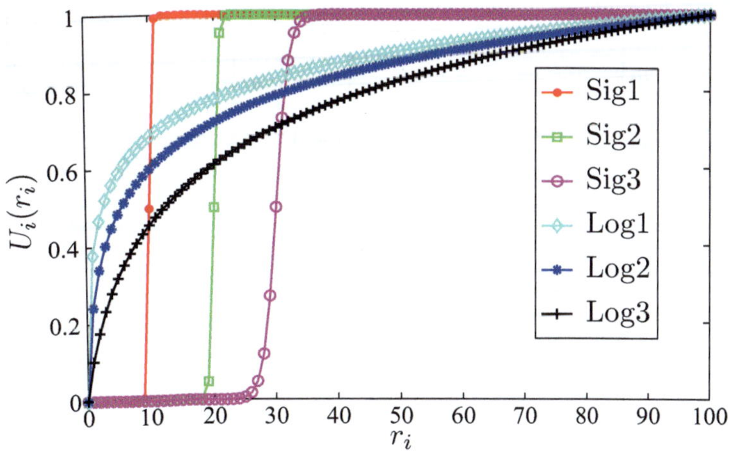

Fig. 4.18 The users utility functions $U_i(r_i)$ used in the simulation (three sigmoidal-like functions and three logarithmic functions)

Table 4.1 Users and their applications utilities

Applications utilities parameters		Users indexes
Sig1	Sig $a_i = 5,\ b_i = 10$	$i = \{5\}$
Sig2	Sig $a_i = 3,\ b_i = 20$	$i = \{1\}$
Sig3	Sig $a_i = 1,\ b_i = 30$	$i = \{2, 6\}$
Log1	Log $k_i = 15,\ r_i^{\max} = 100$	$i = \{7\}$
Log2	Log $k_i = 3,\ r_i^{\max} = 100$	$i = \{3\}$
Log3	Log $k_i = 0.5,\ r_i^{\max} = 100$	$i = \{4, 8\}$

use $k_i = 15$ for UE with index $i = \{7\}$, $k_i = 3$ for UE with index $i = \{3\}$, and $k_i = 0.5$ for UEs with indexes $i = \{4, 8\}$, as shown in Fig. 4.18. A summary is shown in Table 4.1. We use an application minimum required rate that is equivalent to the inflection point of the sigmoidal-like function, i.e., $r_i^{\text{req}} = b_i$, for each VIP user running a real-time application, we use $r_i^{\text{req}} = 15$ for each VIP user running a delay-tolerant application and $r_i^{\text{req}} = 0$ for each regular user whether it is running real-time application or delay-tolerant application.

4.3.4.1 Carrier 1 Allocated Rates

In the following simulations, we set $\delta = 10^{-3}$, carrier 1 rate R_1 takes values between 60 and 150 with step of 10. In Fig. 4.19, we show the allocated rates $r_i^{1,\text{all}}$ of different users with different values of carrier 1 total rate R_1 and observe how the proposed rate allocation algorithm converges for different values of R_1. In Fig. 4.19, we show that both VIP and regular users in user group $\mathscr{M}_1$ are allocated resources by carrier 1 when $60 \leq R_1 \leq 150$ since carrier 1 available resources R_1 is greater than the total applications minimum required rates for users in $\mathscr{M}_1$. Figure 4.19 also shows that by

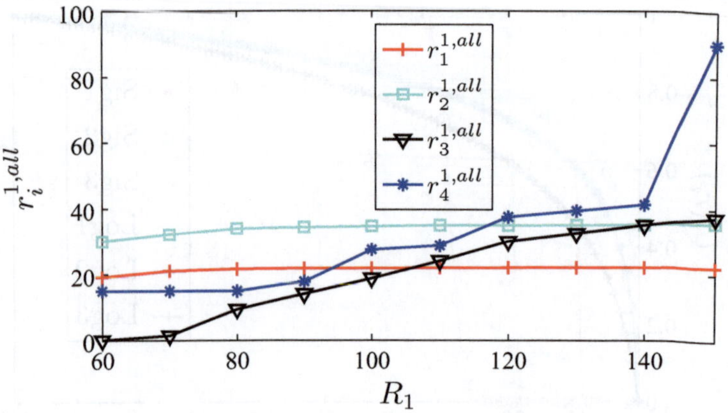

Fig. 4.19 The rates $r_i^{1,\text{all}}$ allocated from carrier 1 to $\mathcal{M}_1$ user group with carrier 1 available resources $60 < R_1 < 150$

using the proposed RA with user discrimination algorithm, no user is allocated zero rate (i.e., no user is dropped). However, carrier 1 resources are first allocated to the VIP users until each of their applications reaches the application minimum required rate r_i^{req}. Then the majority of carrier 1 resources are allocated to the UEs running adaptive real-time applications until they reach their inflection rates, the eNodeB then allocates more of carrier 1 resources to UEs with delay-tolerant applications.

4.3.4.2 Carrier 2 Allocated Rates and the Total Aggregated Rates

In the following simulations, we set $\delta = 10^{-3}$, carrier 2 rate R_2 takes values between 10 and 150 with step of 10 and carrier 1 rate is fixed at $R_1 = 60$. In Fig. 4.20, we show the allocated rates $r_i^{2,\text{all}}$ and the final aggregated rates r_i of different users with different values of carrier 2 total rate R_2 and observe how the proposed rate allocation algorithm converges for different values of R_2. In Fig. 4.20a, we show that when $10 \leq R_2 \leq 45$, only VIP users in $\mathcal{M}_2$ (i.e., UEs in $\mathcal{M}_2^{\text{VIP}}$) that were not allocated resources by carrier 1 or did not reach their applications minimum required rates are allocated resources by carrier 2. Whereas when $45 < R_2 \leq 150$, both VIP and regular users in $\mathcal{M}_2$ are allocated resources by carrier 2 as carrier 2 total rate R_2 is greater than $\sum_{i=1}^{M_2^{\text{VIP}}} q_i^2$ (i.e., the total required rates for UEs to reach their r_i^{req}). Figure 4.20a also shows that by using the proposed RA with user discrimination algorithm that is based on carrier aggregation, the eNodeB takes into consideration the rates allocated to users in $\mathcal{M}_2$ by carrier 1 when allocating carrier 2 resources. Carrier 2 resources are first allocated to VIP users until each of their applications reaches the application minimum required rate r_i^{req}. Then the majority of carrier 2 resources are allocated to the UEs running adaptive real-time applications until they reach their inflection rates, the eNodeB then allocates more of carrier 2 resources to UEs with delay-tolerant applications.

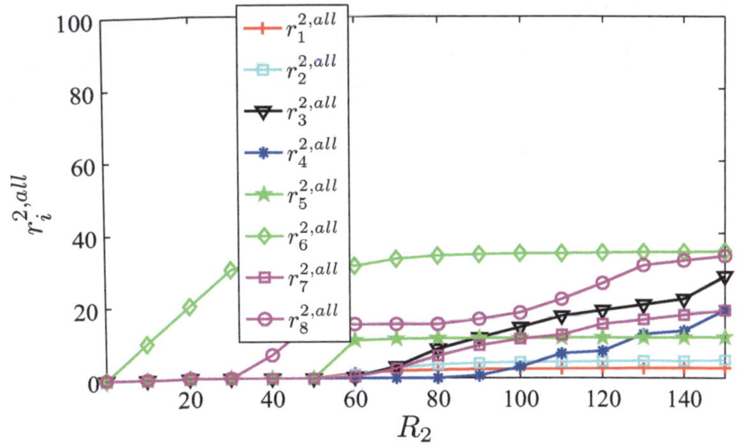

(a)

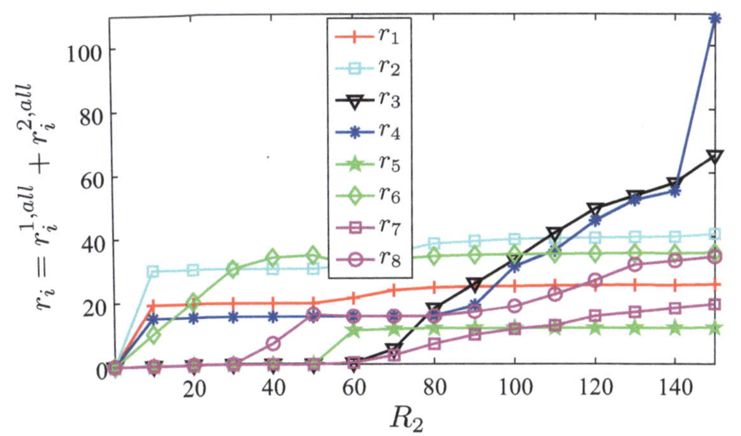

(b)

Fig. 4.20 The rates $r_i^{2,\text{all}}$ allocated from carrier 2 to users in $\mathscr{M}_2$ and the total aggregated rates allocated to the eight users with carrier 2 available resources $10 < R_2 < 150$ and carrier 1 resources fixed at $R_1 = 60$. (**a**) The rates $r_i^{2,\text{all}}$ allocated from carrier 2 to $\mathscr{M}_2$ user group. (**b**) The total aggregated rates r_i allocated by the eNodeB to the eight users

Figure 4.20b shows the total aggregated rates $r_i = \sum_{j=1}^{2} r_i^{j,\text{all}}$ for the eight users.

4.3.4.3 Pricing Analysis for Carrier 1 and Carrier 2

In the following simulations, we set $\delta = 10^{-3}$. In Fig. 4.21, we show carrier 1 shadow price with $60 \leq R_1 \leq 150$. We observe that carrier 1 price p^1 is traffic-dependent as it decreases for higher values of R_1. In Fig. 4.22, we show the offered price of carrier 2 with $10 \leq R_2 \leq 150$ and $R_1 = 60$. We observe that p^2 decreases

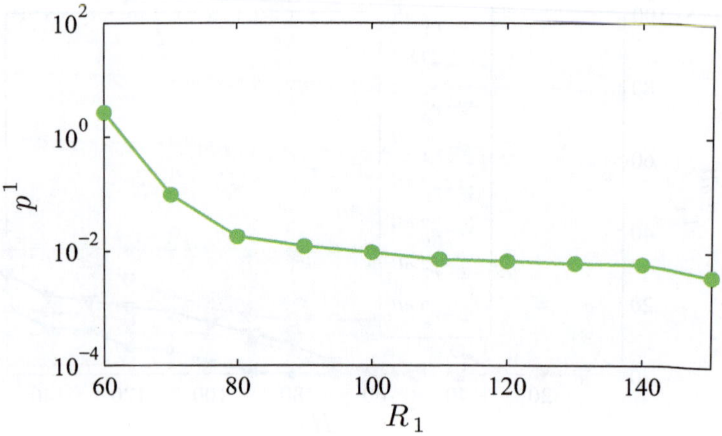

Fig. 4.21 Carrier 1 shadow price p^1 with carrier 1 resources $60 < R_1 < 150$

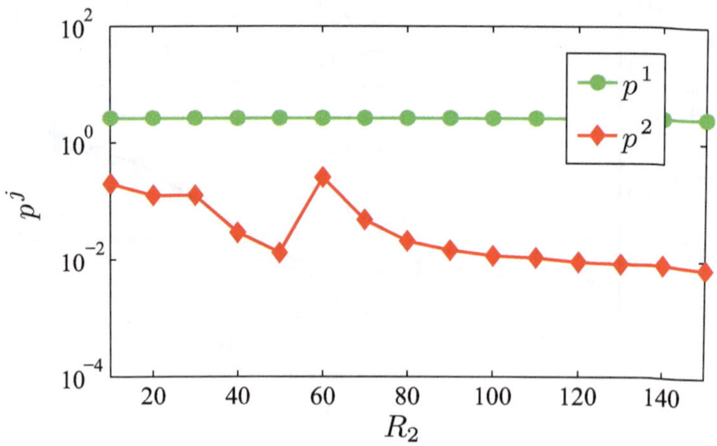

Fig. 4.22 Carrier 1 shadow price p^1 and carrier 2 shadow price p^2 with carrier 2 resources $10 < R_2 < 150$ and carrier 1 resources fixed at $R_1 = 60$

when R_2 increases for $10 \leq R_2 \leq 45$, only VIP users are allocated rates by carrier 2 when $10 \leq R_2 \leq 45$. However, we observe a jump in the price when $R_2 = 50$ as more users are considered in the rate allocation process (i.e., VIP users and regular users in $\mathscr{M}_2$). Figure 4.22 also shows that carrier 2 price p^2 decreases when R_2 increases for $50 \leq R_2 \leq 150$.

4.4 Summary and Conclusion

In this chapter, we introduced a framework for the problem of resource allocation with user discrimination in cellular systems. In Sect. 4.1, We presented a spectrum sharing approach between two types of users; i.e., public safety users and commercial users, running delay-tolerant or real-time applications. We proposed an iterative decentralized RA algorithm for the eNodeB and both the public safety and commercial UEs. The algorithm provides a utility proportional fair resource allocation which guarantees a minimum QoS based on the public safety UEs application target rates, the group that the UE belongs to and the eNodeB available resources. The public safety users group is given priority over the commercial users group and within each group, users running real-time applications are prioritized over those running delay-tolerant applications. We showed through simulations that our algorithm converges to the optimal rates.

In Sect. 4.2, we proposed a novel RA approach to allocate a single eNodeB resources optimally among multi-application UEs with different priority. Two cases of RA optimization problems are considered in our approach. The two cases are based on the total applications target rates of the VIP UEs compared to the eNodeB available resources. A two-stage RA algorithm is presented for each case to allocate the eNodeB resources among users and their applications. Different parameters are taken into consideration by our algorithm when allocating resources such as the application type, the application target rate (if the user application has one), the user subscription weight, and the application weight. We showed through simulations that our two-stage RA algorithm converges to the optimal rates.

In Sect. 4.3, we proposed an efficient resource allocation with user discrimination approach for 5G systems to allocate multiple carriers resources optimally among UEs that belong to different user groups classes. We used utility functions to represent the applications running on the UEs. Each user is assigned a minimum required application rate based on its class and the type of its application. Users are partitioned into different user groups based on their class and the carriers coverage area. We presented resource allocation optimization problems based on carrier aggregation for different cases. We proved the existence of a tractable global optimal solution. We presented a RA algorithm for allocating resources from different carriers optimally among different classes of mobile users. The proposed algorithm ensures fairness in the utility percentage, gives priority to VIP users and within a VIP or a regular user group it gives priority to adaptive real-time applications while providing a minimum QoS for all users. We showed through simulations that the proposed resource allocation algorithm converges to the optimal rates. We also showed that the pricing provided by our algorithm depends on the traffic load.

4.5 C++ Codes

Some functions in the codes are listed in Appendix.

4.5.1 First Code

```
#include<iostream>
#include<math.h>
#include <string>          //This is for string
#include <windows.h>       //This is for the console code

using namespace std;

const float lamda = 0.01;
float low = 0.01;
float weight_all[8][24];
float rate_all[8][23];
float rate_optimal_all[8][24];
float price_all[23];
float weight_all2[8][18];
float rate_all2[8][18];
float rate_optimal_all2[8][18];
float price_all2[18];
int counter = 0;
//main
int main(){
    SetWindow(100,10000);          //This sets the size of the
        window to Width 100, Height 1000
    cout <<"prog is running...Done! \n"<<endl;
    rateCalc1();
      cout<<endl;
      cout<<endl;
      cout<<" price "<<endl;
     for(int j = 0; j<24; j++){
         cout<<price_all[j]<<" ";
     }
     for(int i = 0; i<4;i++){
         cout<<endl;
         cout<<"user # "<<i<<endl;
         cout<<endl;
         cout<<" weight for user # "<<i<<endl;
         for(int j = 0; j<24; j++){
             cout<<weight_all[i][j]<<" ";
         }
         cout<<endl;
         cout<<" ri(n) for user # "<<i<<endl;
         for(int j = 0; j<24; j++){
             cout<<rate_all[i][j]<<" ";
```

```
            }
            cout<<endl;
            cout<<" optimal for user # "<<i<<endl;
            for(int j = 0; j<24; j++){
                cout<<rate_optimal_all[i][j]<< " ";
            }
        }
        //------------------------------//
    rateCalc2();
            cout<<endl;
            cout<<endl;
            cout<<" price "<<endl;
        for(int j = 0; j<18; j++){
                cout<<price_all2[j]<<" ";
            }
        for(int i = 0; i<8;i++){
            cout<<endl;
            cout<<"user # "<<i<<endl;
            cout<<endl;
            cout<<" weight for user # "<<i<<endl;
            for(int j = 0; j<18; j++){
                cout<<weight_all2[i][j]<<" ";
            }
            cout<<endl;
            cout<<" ri(n) for user # "<<i<<endl;
            for(int j = 0; j<18; j++){
                cout<<rate_all2[i][j]<<" ";
            }
            cout<<endl;
            cout<<" optimal for user # "<<i<<endl;
            for(int j = 0; j<18; j++){
                cout<<rate_optimal_all2[i][j]<< " ";
            }
        }
        system("PAUSE");
        return 0;
}
//------------------------------------------//
//The below function returns the root for the sigmoidal-like
    utility function that represents the ideal rate for the
    given shadow price.
float get_roots1_target(float num1,float num2,float p,float
    opt1,float opt2,float target,float R){
        float mid;
        float val;
        float low=.001;
        float high=R;
        for(int i = 0; i < 1000; ++i){
                mid = (low + high)/2;
                float val = F1(mid,num1,num2,p,opt1,opt2);
                if(val < 0.001 && val > -0.001){
                break;
                }
                else if (val < 0 )
```

```
                    high = mid;
                else
                    low = mid;
        }
        if (mid<target){
            mid=mid;
        }
        else
            mid=target;
        return mid;
}
//--------------------------//
//The below function returns the root for the logarithmic
    utility function that represents the ideal rate for the
    given shadow price.
float get_roots2_target(float num,float p,float opt1,float
    opt2,float target,float R){
        float mid;
        float val;
        float low=.001;
        float high=R;
        for(int i = 0; i < 1000; ++i){
                mid = ((low + high)/2);
                float val = F2(mid,num,p,opt1,opt2);
                if(val < 0.001 && val > -0.001){
                        break;
                }
                else if (val < 0 )
                        high = mid;
                else
                        low = mid;
        }
        if (mid<target){
            mid=mid;
        }
        else
            mid=target;
        return mid;
}
//----------------------------------------------------------//
//For the First Case
void rateCalc1(void){
    float weight[4] ={10,10,10,10};
    float prev_Weight[4] = {0,0,0,0};
    float p_n;
    float p_prev = 0;
    float k_i[2] = {3,.5}; //paramethers for the logarithmic
        utility functions. One for each user
    float a_i[2] = {3,1}; //Parameters for the sigmoidal-like
         utility functions
    float b_i[2] = {20,30}; //Parameters for the sigmoidal-
        like utility functions
```

```
    float opt1[8] = {0,0,0,0,0,0,0,0}; //The constant opt is zero for the commercial users and the public safety users.
    float opt2[8] = {0,0,0,0,0,0,0,0};
    float rateArray[8]={0,0,0,0,0,0,0,0};
    float target[4]={20,30,15,15}; // the application target rate for each public safety user
    float Rate = 70;
    int N_users = 4;
    float r_i_optimal[8];
    int counter = 0;

    p_n = shadowPrice(weight,Rate,N_users);

    //Checking whether the difference between the current bid and the previous one is less than a threshold Δ for all users.
    while (((abs(weight[0] - prev_Weight[0]) ≥ lamda) || (abs(weight[1] - prev_Weight[1]) ≥ lamda) || (abs(weight[2] - prev_Weight[2]) ≥ lamda) || (abs(weight[3] - prev_Weight[3]) ≥ lamda))){

        price_all[counter] = p_n;

        //printing  weight:

        for(int i=0; i<4; i++){
            weight_all[i][counter] = weight[i];
        }

        //Finding r_i_optimal
        for(int i = 0; i <4; i++){
            r_i_optimal[i] = weight[i]/p_n;
            rate_optimal_all[i][counter] = r_i_optimal[i];
        }
        //Finding the ideal rates for users with the public safety sigmoidal-like utility functions
        for(int j = 0; j<2; j++){
            rateArray[j] = get_roots1_target(a_i[j],b_i[j],p_n,opt1[j],opt2[j],target[j],Rate);
            rate_all[j][counter] = rateArray[j];
        }
        //finding the ideal rates for users with public safety logarithmic utility functions
        for(int j=2; j<4; j++){
            rateArray[j] = get_roots2_target(k_i[j-2],p_n,opt1[j],opt2[j],target[j],Rate);
            rate_all[j][counter] = rateArray[j];
        }

        // Giving values to Weight:
        for(int k = 0; k<4; k++){
            prev_Weight[k] = weight[k];
        }
```

```
        //Finding new weights
        for(int k = 0; k<4; k++){
            weight[k] = p_n * rateArray[k];
        }
        p_prev = p_n;
        p_n = shadowPrice(weight,Rate,N_users);
        counter++;
    }
}
//-------------------------------------------------------//
//For the Second Case
void rateCalc2(void){
    float weight[8] ={10,10,10,10,10,10,10,10};
    float prev_Weight[8] = {0,0,0,0,0,0,0,0};
    float p_n;
    float p_prev = 0;
    float k_i[2] = {3,.5}; //paramethers for the logarithmic
        utility functions. One for each user
    float a_i[2] = {3,1}; //Parameters for the sigmoidal-like
        utility functions
    float b_i[2] = {20,30}; //Parameters for the sigmoidal-
        like utility functions
    float opt1[8] = {20,30,15,15,0,0,0,0}; //The constant opt
         is zero for the commercial users since they don't
        send application target rates and is equal to the
        application target rate for public safety users.
    float opt2[8] = {0,0,0,0,0,0,0,0};
    float rateArray[8]={0,0,0,0,0,0,0,0};
    float Rate = 200;
    int N_users = 8;
    float r_i_optimal[8];
    int counter=0;

    p_n = shadowPrice(weight,Rate,N_users);

    //Checking whether the difference between the current bid
         and the previous one is less than a threshold ∆ for
        all users.
    while (((abs(weight[0] - prev_Weight[0]) ≥ lamda) || (abs
        (weight[1] - prev_Weight[1]) ≥ lamda) || (abs(weight
        [2] - prev_Weight[2]) ≥ lamda) || (abs(weight[3] -
        prev_Weight[3]) ≥ lamda) || (abs(weight[4] -
        prev_Weight[4]) ≥ lamda) || (abs(weight[5] -
        prev_Weight[5]) ≥ lamda) || (abs(weight[6] -
        prev_Weight[6]) ≥ lamda) || (abs(weight[7] -
        prev_Weight[7]) ≥ lamda))){

        price_all2[counter] = p_n;

        for(int i=0; i<8; i++){
            weight_all2[i][counter] = weight[i];
        }

        //Finding r_i_optimal
```

```
        for(int i = 0; i <8; i++){
            r_i_optimal[i] = weight[i]/p_n;
            rate_optimal_all2[i][counter] = r_i_optimal[i];
        }
        //Finding the ideal rates for users with the public safety sigmoidal-like utility functions
        for(int j = 0; j<2; j++){
            rateArray[j] = get_roots1(a_i[j],b_i[j],p_n,opt1[j],opt2[j],Rate)+opt1[j];
            cout<<  rateArray[j]<<" #"<<j<<"     *** rate"<<endl;
            rate_all2[j][counter] = rateArray[j];
        }
        //finding the ideal rates for users with public safety logarithmic utility functions
        for(int j=2; j<4; j++){
            rateArray[j] = get_roots2(k_i[j-2],p_n,opt1[j],opt2[j],Rate)+opt1[j];
            cout<<  rateArray[j]<<" #"<<j<<"     *** rate"<<endl;
            rate_all2[j][counter] = rateArray[j];
        }
        //finding the ideal rates for users with a commercial sigmoidal-like utility functions
        for(int j = 4; j<6; j++){
            rateArray[j] = get_roots1(a_i[j-4],b_i[j-4],p_n,opt1[j],opt2[j],Rate)+opt1[j];
            cout<<  rateArray[j]<<" #"<<j<<"     *** rate"<<endl;
            rate_all2[j][counter] = rateArray[j];
        }
        //finding the ideal rates for users with a logarithmic utility functions
        for(int j=6; j<8; j++){
            rateArray[j] = get_roots2(k_i[j-6],p_n,opt1[j],opt2[j],Rate)+opt1[j];
            cout<<  rateArray[j]<<" #"<<j<<"     *** rate"<<endl;
            rate_all2[j][counter] = rateArray[j];
        }
        //Giving values to weight:
        for(int k = 0; k<8; k++){
            prev_Weight[k] = weight[k];
        }
        for(int k = 0; k<8; k++){
            weight[k] = p_n * rateArray[k];
        }

        p_prev = p_n;
        p_n = shadowPrice(weight,Rate,N_users);
        counter++;
    }
}
```

4.5.2 Second Code

```
#include<iostream>
#include<math.h>
#include <string>        //This is for string
#include <windows.h>     //This is for the console code

using namespace std;

float c_i[3];
const float lamda = 0.001;
float low = 0.001;
float weight_all[8][24];
float rate_all[8][23];
float rate_optimal_all[2][40];
float rate_optimal_all2[2][40];
float price_all[23];
int counter = 0;
float rate_all1[2][40];
float rate_all2[2][40];
float rate_all3[2][40];
float rate_all2_R[2][10];
float rate_allR_S1[2][10];
float rate_optimal_all_S1[2][10];
//----------------------------------------//
//main
int main(){
    SetWindow(100,10000);         //This sets the size of the
        window to Width 100, Height 1000
    cout <<"prog is running...Done! \n"<<endl;

    cout<<" First Stage-RA to UEs "<<endl;
        for (int i=1; i<11;i++){
        int time= i-1;
             float Rate=i*5;
        rateCalc1(Rate,time);
    }
    cout<<endl;
    cout<<endl;
    for(int i = 0; i<2; i++){
    cout<<endl;
        cout<<" rate1 ri(n) for user # "<<i<<endl;
        for(int j = 0; j<10; j++){
            cout<<rate_allR_S1[i][j]<<" ";
        }
            cout<<endl;
        cout<<" optimal rate1 for user # "<<i<<endl;
        for(int j = 0; j<10; j++){
            cout<<rate_optimal_all_S1[i][j]<<" ";
        }
    }
```

```
    //------------------------------------------------//
    cout<<endl;
    cout<<endl;
    cout<<" Second Stage-Internal UE RA "<<endl;
    float opt1[2]={20,30};
    float opt2[2]={0,0};
    float rate
        [10]={4.09382,9.08366,14.0799,19.0006,19.8612,19.8693,
            19.8671, 19.8676,19.8752,20.4029};
    float ratee
        [10]={0.90618,0.916342,0.920064,0.99943,5.13875,
            10.1307,15.1329, 20.1324,25.1248,29.5971};
    float k_i[2] = {3,.5};
    float a_i[2] = {3,1};
    float b_i[2] = {20,30};
    float a1[2]={0.5,0.9};
    float a2[2]={0.5,0.1};
    int time2=0;

    cout<<endl;
        for (int i=0; i<1; i++){
        cout<<endl;
        cout<<"user # "<<i<<endl;
        cout<<endl;
        for (int n=0;n<10;n++){
        float rate1=rate[n];
        int time2=n;
        rateCalc2(time2,rate1,opt1[i],opt2[i],k_i[i],a_i[i],
            b_i[i],a1[i],a2[i]);
        cout<<endl;
        }
        for(int k=0;k<2;k++){
        cout<<endl;
        cout<<" rate2 rij(n) for user 1 application # "<<k<<
            endl;
        for(int j = 0; j<10; j++){
        cout<<rate_all2_R[k][j]<<" ";
        }
        }
        }
        for (int i=1; i<2; i++){
        cout<<endl;
        cout<<endl;
        cout<<"user # "<<i<<endl;
        cout<<endl;
        for (int n=0;n<10;n++){
        float rate2=ratee[n];
        int time2=n;
        rateCalc2(time2,rate2,opt1[i],opt2[i],k_i[i],a_i[i],
            b_i[i],a1[i],a2[i]);
        cout<<endl;
        }
        for(int k=0;k<2;k++){
        cout<<endl;
```

```
            cout<<" rate2 rij(n) for user 2 and application # "<<
                k<<endl;
            for(int j = 0; j<10; j++){
            cout<<rate_all2_R[k][j]<<" ";
            }
            }
            }

        system("PAUSE");
      return 0;
}
//-----------------------------------------------------------//
void rateCalc1(float Rate,int time){
     float weight[2] ={1,1};
     float prev_Weight[2] = {0,0};
     float p_n;
     float p2[2]={0,0};
     float R=Rate;
     float p_prev = 0;
     float k_i[2] = {3,0.5};
     float a_i[2] = {3,1};
     float b_i[2] = {20,30};
     float opt1[2] = {20,30};
     float opt2[2] = {0,0};
     float Alpha1[2] = {0.5,0.9};
     float Alpha2[2] = {0.5,0.1};
     float B[2]={1,1};
     float rateArray[2]={0,0};
     float target[2]={20,30};
     float weight2[2]={0,0};
     int counter = 0;
     int N_users = 2;
     float r_i_optimal[2];

     p_n = shadowPrice(weight,R,N_users);

     while (((abs(weight[0] - prev_Weight[0]) ≥ lamda) || (abs
         (weight[1] - prev_Weight[1]) ≥ lamda))){

         //Finding ri
         for(int i = 0; i <2; i++){
             r_i_optimal[i] = weight[i]/p_n;
         }

         for(int j=0; j<2; j++){
 rateArray[j] = get_roots3_disc(k_i[j],a_i[j],b_i[j],p_n,opt1[
     j],opt2[j],Alpha1[j],Alpha2[j],R,B[j]);
         }

         //Giving values to Weight:
         for(int k = 0; k<2; k++){
             prev_Weight[k] = weight[k];
         }

```

```
        //Finding new weights
        for(int k = 0; k<2; k++){
            weight2[k] = p_n * rateArray[k];
        }

            for(int k = 0; k<2; k++){
            float exp = 2.718281828459;

            float Del_W= 5*pow(exp, (-counter/10));
            if (abs(weight2[k]-prev_Weight[k])>Del_W){
            weight[k] = prev_Weight[k]+(((weight2[k] -
                prev_Weight[k])/abs(weight2[k]-prev_Weight[k])
                )*Del_W);
            }
            else
            weight[k]=weight2[k];
            }
        p_prev = p_n;
        p_n = shadowPrice(weight,R,N_users);
        counter++;
    }
        for(int j=0; j<2; j++){

            cout<< R<<"      *** rate R"<<endl;
            rate_allR_S1[j][time] = rateArray[j];
            rate_optimal_all_S1[j][time] = r_i_optimal[j];
        }
}
//-------------------------------------//
void rateCalc2(int time2,float rate,float opt1,float opt2,
    float k_i,float a_i,float b_i,float a1,float a2){
    float weight[2] ={10,10};
    float prev_Weight[2] = {0,0};
    float p_n=0;
    float p_prev = 0;
    float rateArray[2]={0,0};
    float weight2[2];
    int counter = 0;
    float Ra = rate;
    int N_users = 2;
    float r_i_optimal[2];

    p_n = shadowPrice(weight,Ra,N_users);

    while (((abs(weight[0] - prev_Weight[0]) ≥ lamda) || (abs
        (weight[1] - prev_Weight[1]) ≥ lamda))){

        if (counter≥40){
            break;
        }
        //Finding ri
        for(int i = 0; i <2; i++){
            r_i_optimal[i] = weight[i]/p_n;
        }
```

```

        for(int j=0; j<1; j++){
            rateArray[j] = get_roots1_disc(a_i,b_i,p_n,opt1,
                Ra,a1);
        }
        for(int j=1; j<2; j++){
            rateArray[j] = get_roots2_disc(k_i,p_n,opt2,Ra,a2
                );
        }

        //Giving values to Weight:
        for(int k = 0; k<2; k++){
            prev_Weight[k] = weight[k];
        }

        //Finding new weights
        for(int k = 0; k<2; k++){
            weight2[k] = p_n * rateArray[k];
        }
            for(int k = 0; k<2; k++){
            float exp = 2.718281828459;
            float Del_W= 5*pow(exp, (-counter/10));
            if (abs(weight2[k]-prev_Weight[k])>Del_W){
            weight[k] = prev_Weight[k]+(((weight2[k] -
                prev_Weight[k])/abs(weight2[k]-prev_Weight[k])
                )*Del_W);
            }
            else
                weight[k]=weight2[k];
        }

        p_prev = p_n;
        p_n = shadowPrice(weight,Ra,N_users);
        counter++;
    }
    for(int i=0;i<2;i++){
        rate_all2_R[i][time2]=rateArray[i];
    }
}
```

4.5.3 Third Code

```
#include<iostream>
#include<math.h>
#include <string>        //This is for string
#include <windows.h>     //This is for the console code

using namespace std;
```

```
const float lamda = 0.001;
float low = 0.01;
float rate_all1[4][60];
float rate_all2[4][40];
float rate_optimal_all2[2][40];
float rate_optimal_all[4][60];
float rate_optimal_all_R[4][30];
float rate_optimal_all2_R[2][30];
int counter = 0;
//---------------------------------------//
//main
int main(){
    SetWindow(100,10000);       //This sets the size of the
        window to Width 100, Height 1000
    cout <<"prog is running...Done! \n"<<endl;

    cout<<endl;
    cout<<endl;
    float opt1[4]={20,30,0,0};
    float opt2[4]={0,0,0,0};
    float k_i[4] = {3,.5,3,0.5};
    float a_i[4] = {3,1,3,1};
    float b_i[4] = {20,30,20,30};
    float a1[4]={0.5,0.9,0.5,0.9};
    float a2[4]={0.5,0.1,0.5,1};
    float rate
        [30]={55,60,65,70,75,80,85,90,95,100,105,110,115,120,
                125,130,135,140,145,150,155,160,165,170,175,
                180,185,190,195,200};
    int time=0;

    for (int i=0; i<4; i++){
        cout<<endl;
        cout<<"user # "<<i<<endl;
        cout<<endl;
        for(int j=0;j<30;j++){
        float rate1=rate[j];
        int time=j;
     rateCalc1(time,rate1);
        }
     cout<<endl;
         cout<<" optimal rate ri for user # "<<i<<endl;
         for(int n = 0; n<30; n++){
        cout<<rate_optimal_all_R[i][n]<<" ";
         }
     }
     cout<<endl;
     cout<<endl;
     float rateri1
         [30]={20.001,20.0007,20.0015,20.0015,20.0022,20.001,
             19.9995,20.0016,20.0015,20.2079,20.7419,21.3056,
                 22.1623,26.7664,27.91,27.9764,29.9408,33.6165,
```

```
            37.3947,40.5558,44.0232,47.0263,50.2558,55.525,
          57.6471,61.3589,65.9433,67.6665,68.4792,70.2581};
    float rateri2
        [30]={30.0172,30.0002,30.0017,30.0018,30.0028,30.0178,
            29.9984,30.0019,30.0017,29.9416,31.763,33.6833,
            35.4113,35.4288,35.9849,35.9028,35.6817,35.7717,
            35.48,35.78,36.0042,36.4028,36.7514,36.4735,
            36.79,37.0462,37.0547,37.5147,35.9521,36.3843};
    float rateri3
        [30]={4.07615,9.08453,14.0774,18.998,19.862,19.8689,
            19.8653,19.8674,19.8751,20.2074,20.7401,21.2993,
            22.1523,22.3763,25.336,30.7621,33.6376,
            34.7628,36.0627,37.4167,38.6955,40.1683,
            41.5961,42.3058,43.6951,45.0178,46.0361,
            47.6174,54.1498,55.7189};
    float rateri4
        [30]={0.905678,0.914666,0.919434,0.998632,5.13307,
            10.1123,15.1368,20.1292,25.1217,29.6431,31.755,
            33.7118,35.2742,35.4285,35.7691,35.3586,35.7399,
            35.8491,36.0627,36.2475,36.2771,36.4026,36.3967,
            35.6957,36.8679,36.5771,35.9659,37.2013,
            36.4189,37.6387};
        for (int i=0; i<1; i++){

        cout<<"user # "<<i<<endl;
        cout<<endl;
        for (int n=0;n<30;n++){
        float rate2=rateri1[n];
        int time2=n;
    rateCalc2(time2,rate2,opt1[i],opt2[i],k_i[i],a_i[i],b_i[i
        ],a1[i],a2[i]);
    cout<<endl;
        }
    for(int k=0;k<2;k++){
        cout<<endl;
        cout<<" rate2 rij(n) for user#1 application # "<<k<<
            endl;
         for(int j = 0; j<30; j++){

        cout<<rate_optimal_all2_R[k][j]<<" ";

         }
    }
        }
        cout<<endl;
        for (int i=1; i<2; i++){
        cout<<"user # "<<i<<endl;
        cout<<endl;
        for (int n=0;n<30;n++){
```

```
        float rate2=rateri2[n];
        int time2=n;
    rateCalc2(time2,rate2,opt1[i],opt2[i],k_i[i],a_i[i],b_i[i
        ],a1[i],a2[i]);
    cout<<endl;
        }
    for(int k=0;k<2;k++){
        cout<<endl;
        cout<<" rate2 rij(n) for user#2 application # "<<k<<
            endl;
         for(int j = 0; j<30; j++){
        cout<<rate_optimal_all2_R[k][j]<<" ";
         }
    }
        }
        cout<<endl;
        for (int i=2; i<3; i++){
        cout<<"user # "<<i<<endl;
        cout<<endl;
        for (int n=0;n<30;n++){
        float rate2=rateri3[n];
        int time2=n;
    rateCalc2(time2,rate2,opt1[i],opt2[i],k_i[i],a_i[i],b_i[i
        ],a1[i],a2[i]);
    cout<<endl;
        }
    for(int k=0;k<2;k++){
        cout<<endl;
        cout<<" rate2 rij(n) for user#3 application # "<<k<<
            endl;
         for(int j = 0; j<30; j++){

        cout<<rate_optimal_all2_R[k][j]<<" ";

         }
    }
        }

        cout<<endl;
        for (int i=3; i<4; i++){
        cout<<"user # "<<i<<endl;
        cout<<endl;
        for (int n=0;n<30;n++){
        float rate2=rateri4[n];
        int time2=n;
    rateCalc2(time2,rate2,opt1[i],opt2[i],k_i[i],a_i[i],b_i[i
        ],a1[i],a2[i]);
    cout<<endl;
        }
    for(int k=0;k<2;k++){
        cout<<endl;
        cout<<" rate2 rij(n) for user#4 application # "<<k<<
            endl;
         for(int j = 0; j<30; j++){
```

```
            cout<<rate_optimal_all2_R[k][j]<<" ";
            }
    }
        }
      system("PAUSE");
     return 0;
}
//-----------------------------------------------//
void rateCalc1(int time, float rate1){
     float weight[4] ={10,10,10,10};
     float prev_Weight[4] = {0,0,0,0};
     float p_n;
     float p_prev = 0;
     float k_i[4] = {3,.5,3,.5};
     float a_i[4] = {3,1,3,1};
     float b_i[4] = {20,30,20,30};
     float opt1[4] = {20,30,0,0};
     float opt2[4]={0,0,0,0};
     float Alpha1[4] = {0.5,0.9,0.5,0.9};
     float Alpha2[4] = {0.5,0.1,0.5,0.1};
     float rateArray[4]={0,0,0,0};
     float R=rate1;
     float weight2[4]={0,0,0,0};
     float B[4]={1,1,1,1};
     float r_i_optimal[4];
     int N_users = 4;
     int counter=0;

     p_n = shadowPrice(weight,R,N_users);

     while (((abs(weight[0] - prev_Weight[0]) ≥ lamda) || (abs
          (weight[1] - prev_Weight[1]) ≥ lamda) || (abs(weight
          [2] - prev_Weight[2]) ≥ lamda) || (abs(weight[3] -
          prev_Weight[3]) ≥ lamda))){

          //Finding ri
          for(int i = 0; i <4; i++){
               r_i_optimal[i] = weight[i]/p_n;
          }
          //Finding ri_optimal
          for(int j = 0; j<4; j++){
 rateArray[j] = get_roots3_disc(k_i[j],a_i[j],b_i[j],p_n,opt1[
     j],opt2[j],Alpha1[j],Alpha2[j],R,B[j]);
          }
     for(int k = 0; k<4; k++){
               prev_Weight[k] = weight[k];
          }

          //Giving values to Weight:
          for(int k = 0; k<4; k++){
               weight2[k] = p_n * rateArray[k];
          }

          for(int k = 0; k<4; k++){
```

```
            float exp = 2.718281828459;
            float Del_W= 5*pow(exp, (-counter/10));
            if (abs(weight2[k]-prev_Weight[k])>Del_W){
            weight[k] = prev_Weight[k]+(((weight2[k] -
                prev_Weight[k])/abs(weight2[k]-prev_Weight[k])
                )*Del_W);
            }
            else
                weight[k]=weight2[k];
        }
        p_prev = p_n;
        p_n = shadowPrice(weight,R,N_users);
        counter++;
    }
    for(int i=0;i<4;i++){
        rate_optimal_all_R[i][time]=r_i_optimal[i];
    }
}
//------------------------------------------------------//
//UE internal applications RA
void rateCalc2(int time2,float rate,float opt1,float opt2,
    float k_i,float a_i,float b_i,float a1,float a2){
    float weight[2] ={1,1};
    float prev_Weight[2] = {0,0};
    float p_n=0;
    float p_prev = 0;
    float rateArray[2]={0,0};
    float weight2[2]={0,0};
    float r_i_optimal2[2];
    int counter = 0;
    int N_users = 2;
    float R = rate;

    p_n = shadowPrice(weight,R,N_users);

    while (((abs(weight[0] - prev_Weight[0]) ≥ lamda) || (abs
        (weight[1] - prev_Weight[1]) ≥ lamda))){ //|| (abs(
        weight[2] - prev_Weight[2]) ≥ lamda) || (abs(weight[3]
         - prev_Weight[3]) ≥ lamda))){

        //Finding ri
        for(int i = 0; i <2; i++){
            r_i_optimal2[i] = weight[i]/p_n;
        }

        for(int j=0; j<1; j++){
            rateArray[j] = get_roots1_disc(a_i,b_i,p_n,opt1,R
                ,a1);
        }
        for(int j=1; j<2; j++){
            rateArray[j] = get_roots2_disc(k_i,p_n,opt2,R,a2)
                ;
        }

```

```
        //Giving values to Weight:
        for(int k = 0; k<2; k++){
        prev_Weight[k] = weight[k];
        }
        //Finding new weights
            for(int k = 0; k<2; k++){
            weight2[k] = p_n * rateArray[k];
        }

        for(int k = 0; k<2; k++){
            float exp = 2.718281828459;
            float Del_W= 5*pow(exp, (-counter/10));
            if (abs(weight2[k]-prev_Weight[k])>Del_W){
            weight[k] = prev_Weight[k]+(((weight2[k] -
                prev_Weight[k])/abs(weight2[k]-prev_Weight[k])
                )*Del_W);
            }
            else
                weight[k]=weight2[k];
    }

        p_prev = p_n;
        p_n = shadowPrice(weight,R,N_users);
        counter++;
    }
    for(int i=0;i<2;i++){
        rate_optimal_all2_R[i][time2]=r_i_optimal2[i];
    }
}
```

4.5.4 Fourth Code

```
#include<iostream>
#include<math.h>
#include <string>       //This is for string
#include <windows.h>    //This is for the console code

using namespace std;
const float lamda = 0.001;
float low = 0.01;
float rate_all1[4][60];
float rate_all2[4][40];
float rate_optimal_all2[2][40];
float rate_optimal_all[4][60];
float rate_optimal_all_R[4][15];
float rate_optimal_all2_R[2][30];
int counter = 0;
float rate_optimal_all_R_C1[4][15];
```

```
float rate_optimal_all_R_C2[8][15];
float carrier1_rates[8];
float price1[15];
float price1C1[15];
float price2[15];
//-----------------------------------------//
//main
int main(){
    SetWindow(100,10000);       //This sets the size of the
        window to Width 100, Height 1000
    cout <<"prog is running...Done! \n"<<endl;
    cout<<endl;
    cout<<endl;
    float k_i[4] = {3,.5,15,0.5};
    float a_i[4] = {3,1,5,1};
    float b_i[4] = {20,30,10,30};
    float rate
        [15]={10,20,30,40,50,60,70,80,90,100,110,120,130,
              140,150};
    int time=0;
    for(int i = 0; i<8; i++){
    for(int n = 0; n<15; n++){
        rate_optimal_all_R_C2[i][n]=0;
    }}
    ////////////////////////////////////
    for (int i=0; i<4; i++){
        cout<<endl;
        cout<<"user # "<<i<<endl;
        cout<<endl;
        for(int j=0;j<15;j++){
        float rate1=60;
        int time=j;
    rateCalc1(time,rate1);
        }

    cout<<endl;
        cout<<" optimal rate ri for user # "<<i<<endl;
        for(int n = 0; n<15; n++){
        cout<<rate_optimal_all_R[i][n]<<" ";
        }
    }
        cout<<endl;
        cout<<" Carrier1 prices "<<endl;
        for(int n = 0; n<15; n++){
        cout<<price1[n]<<" ";
         }
    cout<<endl;
    cout<<endl;
    float carrier1_rates[8]={0,0,0,0,0,0,0,0};
    for(int i = 0; i<4; i++){
        carrier1_rates[i]=rate_optimal_all_R[i][0];
    }
    ///////////////////////////////////////////
    cout<<endl;
```

```
cout<<endl;
 for(int i = 0; i<4; i++){
 for(int n = 0; n<15; n++){

    rate_optimal_all_R_C1[i][n]=0;
 }}
 for(int n = 0; n<15; n++){
    price1C1[n]=0;
 }
    for (int i=0; i<4; i++){
    cout<<endl;
    cout<<"user # "<<i<<endl;
    cout<<endl;
    for(int j=5;j<15;j++){
    float rate1=rate[j];
    int time=j;
rateCalc1(time,rate1);
    }
cout<<endl;
     cout<<" Carrier 1_Different R1 optimal rate ri for
         user # "<<i<<endl;
     for(int n = 5; n<15; n++){
    cout<<rate_optimal_all_R_C1[i][n]<<" ";
     }
     }
    cout<<endl;
    cout<<endl;
    cout<<" Carrier1 prices for Different R1 "<<endl;
    for(int n = 5; n<15; n++){
    cout<<price1C1[n]<<" ";
     }
cout<<endl;
cout<<endl;
/////////////////////////////////////////
cout<<"Carrier 2"<<endl;
    for (int i=0; i<8; i++){
    cout<<endl;
    cout<<"user # "<<i<<endl;
    cout<<endl;
    for(int j=0;j<5;j++){
    float rate2=rate[j];
    int time=j;
rateCalc2(time,rate2);
    }
cout<<endl;
    cout<<" optimal rate ri from carrier 2 for user # "<<
        i<<endl;
    for(int n = 0; n<5; n++){
    cout<<rate_optimal_all_R_C2[i][n]<<" ";
    }
}
cout<<endl;
cout<<endl;
/////////////////////////////////////////////
```

```
    for (int i=0; i<8; i++){
        cout<<endl;
        cout<<"user # "<<i<<endl;
        cout<<endl;
        for(int j=5;j<15;j++){
        float rate3=rate[j];
        int time=j;
    rateCalc3(time,rate3,carrier1_rates);
        }
    cout<<endl;
         cout<<" optimal rate ri from carrier 2 for user # "
             <<i<<endl;
         for(int n = 0; n<15; n++){
         cout<<rate_optimal_all_R_C2[i][n]<<" ";
         }
    }
        cout<<endl;
        cout<<endl;
        for(int i = 0; i<8; i++){
        cout<<" Aggregated optimal rate ri from carrier 1 and
             2 for user group 2 # "<<i<<endl;
        for(int n = 0; n<15; n++){
        cout<<rate_optimal_all_R_C2[i][n]+rate_optimal_all_R[
            i][n]<<" ";
        }
        cout<<endl;
        cout<<endl;
        }
        cout<<endl;
        cout<<" Carrier2 prices "<<endl;
        for(int n = 0; n<15; n++){
        cout<<price2[n]<<" ";
         }
     cout<<endl;
     cout<<endl;
     system("PAUSE");
     return 0;
}
//---------------------------------------------------------//
void rateCalc1(int time, float rate1){
     float weight[4] = {10,10,10,10};
     float prev_Weight[4] = {0,0,0,0};
     float p_n;
     float p_prev = 0;
     float k_i[4] = {3,.5,15,.5};
     float a_i[4] = {3,1,5,1};
     float b_i[4] = {20,30,10,30};
     float opt1[4] = {0,30,0,15};
     float opt2[4] = {0,0,0,0};
     float rateArray[4] = {0,0,0,0};
     float R=rate1;
     float weight2[4] = {0,0,0,0};
     float r_i_optimal[4];
     int N_users = 4;
```

```
    int counter= 0;


    p_n = shadowPrice(weight,R,N_users);

    while (((abs(weight[0] - prev_Weight[0]) ≥ lamda) || (abs
        (weight[1] - prev_Weight[1]) ≥ lamda) || (abs(weight
        [2] - prev_Weight[2]) ≥ lamda) ||(abs(weight[3] -
        prev_Weight[3]) ≥ lamda))){

        //Finding ri
        for(int i = 0; i <4; i++){
            r_i_optimal[i] = weight[i]/p_n;
        }
        //Finding ri_optimal
        for(int j = 0; j<2; j++){
            rateArray[j] = get_roots1(a_i[j],b_i[j],p_n,opt1[
                j],opt2[j],R)+opt1[j];
        }

        for(int j = 2; j<4; j++){
            rateArray[j] = get_roots2(k_i[j-2],p_n,opt1[j],
                opt2[j],R)+opt1[j];
        }

        for(int k = 0; k<4; k++){
            prev_Weight[k] = weight[k];
        }

        //Giving values to Weight:
        for(int k = 0; k<4; k++){
            weight2[k] = p_n * rateArray[k];
        }

        for(int k = 0; k<4; k++){
            float exp = 2.718281828459;
            float Del_W= 5*pow(exp, (-counter/10));
            if (abs(weight2[k]-prev_Weight[k])>Del_W){
            weight[k] = prev_Weight[k]+(((weight2[k] -
                prev_Weight[k])/abs(weight2[k]-prev_Weight[k])
                )*Del_W);
            }
            else
            weight[k]=weight2[k];
        }

        p_prev = p_n;
        p_n = shadowPrice(weight,R,N_users);
        counter++;
    }

    for(int i=0;i<4;i++){
        rate_optimal_all_R[i][time]=rateArray[i];
        rate_optimal_all_R_C1[i][time]=rateArray[i];
```

```
    }
    price1[time]=p_n;
    price1C1[time]=p_n;
}
/////////////////////////////////////////////////////////////
void rateCalc2(int time, float rate2){
    float weight[2] = {1,1};
    float prev_Weight[2] = {0,0};
    float p_n;
    float p_prev = 0;
    float k_i[4] = {3,.5,15,.5};
    float a_i[4] = {3,1,5,1};
    float b_i[4] = {20,30,10,30};
    float opt1[2] = {0,0};
    float opt2[2] = {0,0};
    float rateArray[2] = {0,0};
    float R = rate2;
    float weight2[2] = {0,0};
    float r_i_optimal[2] = {0,0};
    int counter = 0;
    int N_users = 2;

    p_n = shadowPrice(weight,R,N_users);

    while (((abs(weight[0] - prev_Weight[0]) ≥ lamda) || (abs(weight[1] - prev_Weight[1]) ≥ lamda))){

        //Finding ri
        for(int i = 0; i <2; i++){
            r_i_optimal[i] = weight[i]/p_n;
        }
        //Finding ri_optimal
        for(int j = 0; j<1; j++){
            rateArray[j] = get_roots1(a_i[3],b_i[3],p_n,opt1[j],opt2[j],R)+opt1[j];
        }

        for(int j = 1; j<2; j++){
        rateArray[j] = get_roots2(k_i[3],p_n,opt1[j],opt2[j],R)+opt1[j];
        }

        for(int k = 0; k<2; k++){
        prev_Weight[k] = weight[k];
        }

        //Giving values to Weight:
        for(int k = 0; k<2; k++){
            weight2[k] = p_n * rateArray[k];
        }

        for(int k = 0; k<2; k++){
            float exp = 2.718281828459;
            float Del_W= 5*pow(exp, (-counter/10));
```

```
                if (abs(weight2[k]-prev_Weight[k])>Del_W){
                weight[k] = prev_Weight[k]+(((weight2[k] -
                    prev_Weight[k])/abs(weight2[k]-prev_Weight[k])
                    )*Del_W);
                }
                else
                weight[k]=weight2[k];
            }
            p_prev = p_n;
            p_n = shadowPrice(weight,R,N_users);
            counter++;
        }

        float val1 = r_i_optimal[0];
        float val2 = r_i_optimal[1];
        rate_optimal_all_R_C2[5][time]=val1;
        rate_optimal_all_R_C2[7][time]=val2;
        price2[time]=p_n;
}
/////////////////////////////////////////////////////
void rateCalc3(int time, float rate2, float arr_rates[]){
        float weight3[8] = {1,1,1,1,1,1,1,1};
        float prev_Weight3[8] = {0,0,0,0,0,0,0,0};
        float p_n3;
        float p_prev3 = 0;
        float k_i[4] = {3,.5,15,.5};
        float a_i[4] = {3,1,5,1};
        float b_i[4] = {20,30,10,30};
        float opt13[8] = {0,0,0,0,0,30,0,15};
        float opt23[8];
        for(int i=0; i<8; i++){
        opt23[i] = arr_rates[i];
        }
        float rateArray3[8] = {0,0,0,0,0,0,0,0};
        float R3=rate2;
        float weight23[8] = {0,0,0,0,0,0,0,0};
        float r_i_optimal3[8] = {0,0,0,0,0,0,0,0};
        int counter = 0;
        int N_users = 8;

        p_n3 = shadowPrice(weight3,R3,N_users);

        while (((abs(weight3[0] - prev_Weight3[0]) ≥ lamda) || (
            abs(weight3[1] - prev_Weight3[1]) ≥ lamda) || (abs(
            weight3[2] - prev_Weight3[2]) ≥ lamda) || (abs(weight3
            [3] - prev_Weight3[3]) ≥ lamda) || (abs(weight3[4] -
            prev_Weight3[4]) ≥ lamda) || (abs(weight3[5] -
            prev_Weight3[5]) ≥ lamda) || (abs(weight3[6] -
            prev_Weight3[6]) ≥ lamda) || (abs(weight3[7] -
            prev_Weight3[7]) ≥ lamda))){


            //Finding ri
            for(int i = 0; i <8; i++){
```

```
            r_i_optimal3[i] = weight3[i]/p_n3;
        }
        //Finding ri_optimal
        for(int j = 0; j<2; j++){
            rateArray3[j] = get_roots1(a_i[j],b_i[j],p_n3,
                opt13[j],opt23[j],R3)+opt13[j];
        }

        for(int j = 2; j<4; j++){
        rateArray3[j] = get_roots2(k_i[j-2],p_n3,opt13[j],
            opt23[j],R3)+opt13[j];
        }
        for(int j = 4; j<6; j++){
        rateArray3[j] = get_roots1(a_i[j-2],b_i[j-2],p_n3,
            opt13[j],opt23[j],R3)+opt13[j];
        }

        for(int j = 6; j<8; j++){
        rateArray3[j] = get_roots2(k_i[j-4],p_n3,opt13[j],
            opt23[j],R3)+opt13[j];
        }

            for(int k = 0; k<8; k++){
            prev_Weight3[k] = weight3[k];
        }

        //Giving values to Weight:
        for(int k = 0; k<8; k++){
            weight23[k] = p_n3 * rateArray3[k];
        }

        for(int k = 0; k<8; k++){
            float exp = 2.718281828459;
            float Del_W= 5*pow(exp, (-counter/10));
            if (abs(weight23[k]-prev_Weight3[k])>Del_W){
            weight3[k] = prev_Weight3[k]+(((weight23[k] -
                prev_Weight3[k])/abs(weight23[k]-prev_Weight3[
                k]))*Del_W);
            }
            else
            weight3[k]=weight23[k];
        }
        p_prev3 = p_n3;
        p_n3 = shadowPrice(weight3,R3,N_users);
        counter++;
    }
    for(int i = 0; i<8; i++){
    rate_optimal_all_R_C2[i][time]=rateArray3[i];
    }
    price2[time]=p_n3;
}
```

References

1. H. Shajaiah, A. Abdelhadi, C. Clancy, Spectrum sharing between public safety and commercial users in 4g-lte, in *IEEE International Conference on Computing, Networking and Communications (ICNC)* (2014)
2. H. Shajaiah, A. Abdelhadi, C. Clancy, Multi-application resource allocation with users discrimination in cellular networks, in *IEEE International Symposium on Personal, Indoor and Mobile Radio Communications (PIMRC)* (2014)
3. H. Ekstrom, QoS control in the 3GPP evolved packet system. IEEE Commun. Mag. **47**, 76–83 (2009)
4. H. Ekstrom, A. Furuskar, J. Karlsson, M. Meyer, S. Parkvall, J. Torsner, M. Wahlqvist, Technical solutions for the 3G long-term evolution. IEEE Commun. Mag. **44**, 38–45 (2006)
5. I. Research, Mobile VoIP subscribers will near 410 million by 2015; VoLTE still a long way off (Infonetics Research, California, 2010)
6. N. Solutions and Networks, Enhance mobile networks to deliver 1000 times more capacity by 2020 (Nokia Solutions and Networks, 2013)
7. G. Intelligence, Smartphone users spending more 'face time' on apps than voice calls or web browsing (GSMA Intelligence, 2011)
8. N.S. Networks, Understanding smartphone behavior in the network, in *Device Analyzer: Understanding Smartphone Usage*, ed. by D.T. Wagner, A. Rice, A.R. Beresford (Computer Laboratory, University of Cambridge, Cambridge, 2011)
9. Z. Shen, A. Papasakellariou, J. Montojo, D. Gerstenberger, F. Xu, Overview of 3GPP LTE-advanced carrier aggregation for 4G wireless communications. IEEE Commun. Mag. **50**, 122–130 (2012)
10. *Frequency Spectrum Wall Chart* (Commerce Dept., National Telecommunications and Information Administration, Office of Spectrum Management, 2016)
11. S. Shenker, Fundamental design issues for the future internet. IEEE J. Sel. Areas Commun. **13**, 1176–1188 (1995)
12. F. Wilson, I. Wakeman, W. Smith, Quality of service parameters for commercial application of video telephony, in *Proceedings of the Human Factors in Telecommunications Symposium* (Darmstadt, 1993), pp. 139–148
13. G. Tychogiorgos, A. Gkelias, K.K. Leung, Utility-proportional fairness in wireless networks, in *International Symposium on Personal, Indoor, and Mobile Radio Communications* (IEEE, New York, 2012), pp. 839–844
14. A. Abdelhadi, C. Clancy, A utility proportional fairness approach for resource allocation in 4G-LTE, in *IEEE International Conference on Computing, Networking, and Communications (ICNC), CNC Workshop* (2014)
15. A. Abdelhadi, C. Clancy, A robust optimal rate allocation algorithm and pricing policy for hybrid traffic in 4G-LTE, in *IEEE International Symposium on Personal, Indoor, and Mobile Radio Communications (PIMRC)* (2013)
16. Y. Wang, A. Abdelhadi, A QoS-based power allocation for cellular users with different modulations, in *2016 International Conference on Computing, Networking and Communications (ICNC)* (2016), pp. 1–5
17. A. Abdelhadi, H. Shajaiah, Optimal resource allocation for smart phones with multiple applications with MATLAB instructions. Technical Reports (2016)
18. A. Abdelhadi, H. Shajaiah, Application-aware resource allocation with carrier aggregation using MATLAB. Technical Reports (2016)
19. Z. Kbah, A. Abdelhadi, Resource allocation in cellular systems for applications with random parameters, in *2016 International Conference on Computing, Networking and Communications (ICNC)* (2016), pp. 1–5
20. A. Abdelhadi, H. Shajaiah, Optimal resource allocation for cellular networks with MATLAB instructions. CoRR, abs/1612.07862 (2016)

21. J.-W. Lee, R.R. Mazumdar, N.B. Shroff, Downlink power allocation for multi-class wireless systems. IEEE/ACM Trans. Networking **13**, 854–867 (2005)
22. Y. Wang, A. Abdelhadi, T. C. Clancy, Optimal power allocation for LTE users with different modulations, in *2016 Annual IEEE Systems Conference (SysCon)* (2016), pp. 1–5
23. A. Kumar, A. Abdelhadi, T.C. Clancy, A delay efficient multiclass packet scheduler for heterogeneous M2M uplink, in *IEEE MILCOM* (2016)
24. A. Kumar, A. Abdelhadi, T.C. Clancy, An online delay efficient packet scheduler for M2M traffic in industrial automation, in *IEEE Systems Conference* (2016)
25. A. Kumar, A. Abdelhadi, T.C. Clancy, A delay optimal MAC and packet scheduler for heterogeneous M2M uplink. CoRR, abs/1606.06692 (2016)
26. A. Kumar, A. Abdelhadi, T.C. Clancy, A delay-optimal packet scheduler for M2M uplink, in *IEEE MILCOM* (2016)
27. M. Ghorbanzadeh, A. Abdelhadi, C. Clancy, Distributed resource allocation, in *Cellular Communications Systems in Congested Environments* (Springer, Berlin, 2017), pp. 61–91
28. J.B. Taylor, *Principles of Microeconomics*. Microeconomics Series (Houghton Mifflin, Boston, 1998)
29. S. Boyd, L. Vandenberghe, *Introduction to Convex Optimization with Engineering Applications*. Convex Optimization, Course Reader for EE364 (1999)
30. R. Madan, S. Boyd, S. Lall, Fast algorithms for resource allocation in wireless cellular networks. IEEE/ACM Trans. Networking **18**, 973–984 (2010)
31. A. Abdelhadi, C. Clancy, J. Mitola, A resource allocation algorithm for multi-application users in 4G-LTE, in *MobiCom Workshop* (2013)

Chapter 5
Resource Allocation with Carrier Aggregation for Commercial Use of 3.5 GHz Spectrum

The Commission and the President have outlined a path to double the available spectrum for wireless broadband use, and the PCAST Report identifies two technological advances to increase wireless broadband capabilities [1–5]. First, increasing the deployment of small cell networks and second using spectrum sharing technology [6–8]. The 3.5 GHz band is an ideal band for small cell deployments and shared spectrum use because of its smaller coverage. The National Institute of Standards and Technology (NTIA) Fast Track Report [9–12] identified the 3.5 GHz band for potential shared federal and non-federal broadband use. This band is very favorable for commercial cellular systems such as LTE-Advanced systems. Small cells are low-powered wireless base stations designed to play well with macro networks in a heterogeneous network (HetNet). Small cells are backed up by a macro cell layer of coverage so that if a small cell shuts down in the 3.5 GHz shared band, operators can pick up coverage again in the macro network.

In this chapter, we introduce an application-aware spectrum sharing approach for sharing the Federal under-utilized 3.5 GHz spectrum with commercial users. In our model, users are running elastic or inelastic traffic and each application running on the UE is assigned a utility function based on its type. Furthermore, each of the small cells' users has a minimum required target utility for its application. In order for users located under the coverage area of the small cells' eNodeBs, with the 3.5 GHz band resources, to meet their minimum required quality of experience (QoE), the network operator makes a decision regarding the need for sharing the macro cell's resources to obtain additional resources. Our objective is to provide each user with a rate that satisfies its application's minimum required utility through spectrum sharing approach and improve the overall QoE in the network. We present an application-aware spectrum sharing multi-stage algorithm that is based on resource

The content in this chapter is reproduced with permission after modifications (License number 4078221295420). For the original article please refer to [13].

H. Shajaiah et al., *Resource Allocation with Carrier Aggregation in Cellular Networks*, DOI 10.1007/978-3-319-60540-1_5

allocation with carrier aggregation to allocate macro cell permanent resources and small cells' leased resources to UEs based on a utility proportional fairness policy, and allocate each user's application an aggregated rate that can at minimum achieve the application's minimum required utility. The contributions in this chapter are summarized as:

- We present a spectrum sharing approach for sharing the Federal under-utilized 3.5 GHz spectrum with commercial users.
- We present a spectrum sharing algorithm that is based on resource allocation with CA to allocate the small cells' under-utilized 3.5 GHz resources to small cells' users and allocate the macro cell's resources to both macro cell's users and small cell's users that did not reach their applications minimum required utilities by the small cells allocated rates.
- We present simulation results for the performance of the proposed resource allocation algorithm.

5.1 Spectrum Sharing Problem Formulation

We consider LTE-Advanced mobile system consisting of a macro cell, referred to by the index B, with a coverage radius D_B, that is overlaid with S small cells. The macro cell's eNodeB is configured at the LTE-Advanced carrier and the small cell's eNodeB is configured to use the 3.5 GHz under-utilized spectrum band. Let $\mathscr{S}$ denote the set of small cells located within the coverage area of the macro cell B where $S = |\mathscr{S}|$. All small cells are connected to the core network. The small cells are assumed to have a closed access scheme where only registered UEs, referred to by SUEs, are served by the small cells eNodeBs. On the other hand, all UEs under the coverage area of the macro cell B and not within the coverage of any small cell, referred to by MUEs, are served by the macro cell's eNodeB. The set of all MUEs under the coverage area of macro cell B is referred to by μ. The set of SUEs associated to small cell s is referred to by $\mathscr{Q}_s$. We assume that the association of the UEs with their eNodeBs remains fixed during the runtime of the resource allocation process. We have $\bigcup_{s=1}^{S} \mathscr{Q}_s = \Theta$ and $\bigcap_{s=1}^{S} \mathscr{Q}_s = \emptyset$. Each SUE i has a minimum QoE requirement for its applications that is represented by the utility of the user's application with its allocated rate. Let u_i^{req} denote the minimum required utility of SUE $i \in \Theta$.

We express the user satisfaction with its application rates using utility functions [14–23]. We represent the ith user application utility function $U_i(r_i)$ by sigmoidal-like function or logarithmic function where r_i is the rate of the ith user application. Logarithmic utility functions expressed by Eq. (2.2) and sigmoidal-like utility functions expressed by Eq. (2.1) are used to represent delay-tolerant and real-time applications, respectively, as in [14, 23–30].

Figure 5.1 shows a HetNet that consists of one macro cell with one eNodeB and two small cells within the coverage area of the macro cell, each of the small cells has

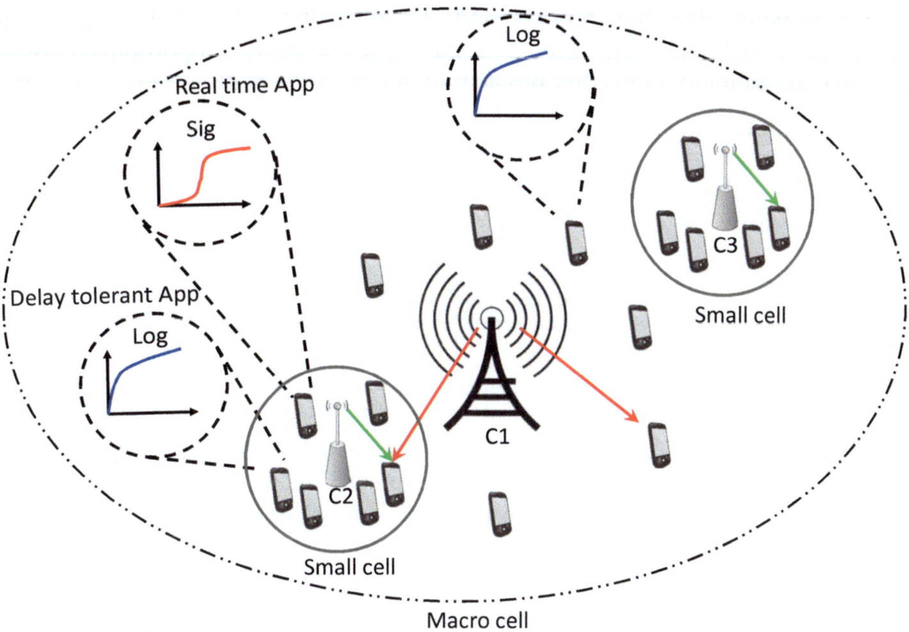

Fig. 5.1 System model for an LTE-Advanced mobile system with one macro cell and two small cells within the coverage area of the macro cell. Each of the small cells is configured to use the 3.5 GHz under-utilized spectrum

one eNodeB that is configured to use the 3.5 GHz under-utilized spectrum. Mobile users under the coverage of the macro cell and the small cells are running real-time or delay-tolerant applications that are represented by sigmoidal-like or logarithmic utility functions, respectively.

5.2 Resource Allocation Optimization for Spectrum Sharing with the 3.5 GHz Spectrum

In this section, we present a resource allocation framework for cellular networks sharing the federal under-utilized 3.5 GHz spectrum. In our model, SUEs are allocated resources from the leased under-utilized 3.5 GHz resources at the small cells eNodeBs whereas MUEs are allocated resources only by the macro cell's eNodeB. Each of the SUEs has a minimum required utility u_i^{req} for each of its applications. First the small cell's eNodeB allocates its available leased resources then the network operator decides which SUEs still require additional resources in order to achieve their minimum required utilities and allocate them more resources from the macro cell eNodeB based on a resource allocation with carrier aggregation optimization problem.

The resource allocation process starts by allocating each of the small cells resources to SUEs under it coverage area. We use a utility proportional fairness resource allocation optimization problem to allocate the small cell resources. The RA optimization problem of the small cell s is given by:

$$\begin{aligned} \max_{\mathbf{r}^s} \quad & \prod_{i=1}^{|\mathcal{Q}_s|} U_i(r_i^s) \\ \text{subject to} \quad & \sum_{i=1}^{|\mathcal{Q}_s|} r_i^s \leq R_s \\ & 0 \leq r_i^s \leq R_s, \qquad i = 1, 2, \ldots, |\mathcal{Q}_s|, \end{aligned} \tag{5.1}$$

where $\mathbf{r}^s = \{r_1^s, r_2^s, \ldots, r_{|\mathcal{Q}_s|}^s\}$, $|\mathcal{Q}_s|$ is the number of SUEs under the coverage area of the small cell s, and R_s is the maximum achievable rate of the under-utilized 3.5 GHz leased spectrum available at the eNodeB of small cell s. The resource allocation objective function is to maximize the entire small cell utility when allocating its resources. It also achieves proportional fairness among utilities such that none of the SUEs will be allocated zero resources. Therefore, a minimum QoS is provided to each SUE. This approach gives real-time applications priority when allocating the small cell resources. The objective function in optimization problem (5.1) is equivalent to $\max_{\mathbf{r}^s} \sum_{i=1}^{|\mathcal{Q}_s|} \log U_i(r_i^s)$. Optimization problem (5.1) is a convex optimization problem and there exists a unique tractable global optimal solution [15].

From optimization problem (5.1), we have the Lagrangian:

$$L_s(\mathbf{r}^s, p^s) = \left(\sum_{i=1}^{|\mathcal{Q}_s|} \log U_i \left(r_i^s\right) \right) - p^s \left(\sum_{i=1}^{|\mathcal{Q}_s|} r_i^s + z_s - R_s \right), \tag{5.2}$$

where $z_s \geq 0$ is the slack variable and p^s is the Lagrange multiplier which is equivalent to the shadow price that corresponds to the service provider's price per unit bandwidth for the small cell resources [15].

The solution of Eq. (5.1) is given by the values r_i^s that solve equation $\frac{\partial \log U_i(r_i^s)}{\partial r_i^s} = p^s$ and are the intersection of the time varying shadow price, horizontal line $y = p^s$, with the curve $y = \frac{\partial \log U_i(r_i^s)}{\partial r_i^s}$ geometrically. Once the RA process is performed by the small cell s, each SUE in $\mathcal{Q}_s$ will be allocated $r_i^{s,\text{all}} = r_i^s$ rate. However, the network operator decides if any of the SUEs requires additional resources in order to reach the minimum required utility u_i^{req} of its application by comparing the utility of the small cell allocated rate that is given by $U_i(r_i^{s,\text{all}})$ with the value u_i^{req}. If the achieved utility for certain SUE is less than the minimum required utility, the network operator requests additional resources from the macro cell for that SUE.

The small cell s eNodeB creates a set $\mathcal{Q}_{sB}$ of all SUEs that needs to be allocated additional resources where $\mathcal{Q}_{sB} = \{\text{SUEs} \in \mathcal{Q}_s \ \ s.t. \ \ u_i^{\text{req}} > U_i(r_i^{s,\text{all}})\}$.

Once each small cell s within the coverage area of the macro cell B performs its RA process based on optimization problem (5.1), the macro cell starts allocating its resources to all MUEs within its coverage area as well as the SUEs that were reported, by the network operator, for their need of additional resources. Let $\mathcal{Q}$ be the set of SUEs that will be allocated additional resources by the macro cell where $\mathcal{Q} = \bigcup_{s=1}^{S} \mathcal{Q}_{sB}$. The set of UEs that will be served by the macro cell's eNodeB, i.e., participate in the macro cell RA process, is given by β where $\beta = \mu \bigcup \mathcal{Q}$. The resource allocation optimization problem of the macro cell B is given by:

$$\begin{aligned} \max_{\mathbf{r}} \quad & \prod_{i=1}^{|\beta|} U_i(r_i + C_i) \\ \text{subject to} \quad & \sum_{i=1}^{|\beta|} r_i \leq R_B \\ & C_i = \begin{cases} 0 & \text{if UE } \ i \notin \mathcal{Q} \\ r_i^{s,\text{all}} & \text{if UE } \ i \in \mathcal{Q} \end{cases} \\ & 0 \leq r_i \leq R_B, \qquad i = 1, 2, \ldots, |\beta|, \end{aligned} \tag{5.3}$$

where $\mathbf{r} = \{r_1, r_2, \ldots, r_{|\beta|}\}$, $|\beta|$ is the number of UEs that will be served by the macro cell's eNodeB, and R_B is the maximum achievable rate of the resources available at the macro cell's eNodeB. The resource allocation objective function is to maximize the entire macro cell utility when allocating its resources. The RA optimization problem (5.3) is based on carrier aggregation. It seeks to maximize the multiplication of the utilities of the rates allocated to MUEs by the macro cell's eNodeB and the utilities of the rates allocated to the SUEs in β by small cells' eNodeBs and macro cell's eNodeB. Utility proportional fairness is used to guarantee that none of the UEs will be allocated zero resources. Real-time applications are given priority when allocating the macro resources using this approach. The objective function in optimization problem (5.3) is equivalent to $\max_{\mathbf{r}} \sum_{i=1}^{|\beta|} \log U_i(r_i + C_i)$. Optimization problem (5.3) is a convex optimization problem and there exists a unique tractable global optimal solution [15].

From optimization problem (5.3), we have the Lagrangian:

$$L_B(\mathbf{r}, p^B) = \left(\sum_{i=1}^{|\beta|} \log U_i(r_i + C_i) \right) - p^B \left(\sum_{i=1}^{|\beta|} r_i + z_B - R_B \right), \tag{5.4}$$

where $z_B \geq 0$ is the slack variable and p^B is the Lagrange multiplier which is equivalent to the shadow price that corresponds to the service provider's price per unit bandwidth for the macro cell resources [15].

The solution of Eq. (5.3) is given by the values r_i that solve equation $\frac{\partial \log U_i(r_i+C_i)}{\partial r_i} = p^B$ and are the intersection of the time varying shadow price, horizontal line $y = p^B$, with the curve $y = \frac{\partial \log U_i(r_i+C_i)}{\partial r_i}$ geometrically. Once the macro cell eNodeB is done performing the RA process based on optimization problem (5.3), each UE in β will be allocated $r_i^{\text{all}} = r_i + C_i$ rate.

5.3 The Macro Cell and Small Cells RA Optimization Algorithm

In this section, we present the macro cell and small cells resource allocation algorithm. The proposed algorithm consists of SUE, MUE, small cell eNodeB, and macro cell eNodeB parts as shown in Algorithms 1–4, respectively. The execution of the algorithm starts by SUEs and MUEs, subscribing for mobile services, transmitting their applications utilities parameters to their corresponding eNodeBs. First, each small cell s eNodeB calculates its allocated rate $r_i^{s,\text{all}}$ to each SUE in $\mathcal{Q}_s$. It then checks whether the achievable utility of that rate is less or greater than the SUE's minimum required utility u_i^{req}. If for any SUE $U_i(r_i^{s,\text{all}}) < u_i^{\text{req}}$, the small cell's eNodeB sends the application parameters and the allocated rate $r_i^{s,\text{all}}$ for that SUE to the macro cell's eNodeB requesting additional resources. Otherwise, it allocates the rate $r_i^{s,\text{all}}$ to that SUE.

Once the macro cell's eNodeB receives the set $\mathcal{Q}_{sB}$ from each small cell in $\mathcal{S}$ within its coverage area, it starts the RA process to allocate its available resources to each UE in β based on a RA with carrier aggregation optimization problem. Once the RA process of the macro cell is performed, the macro cell allocates rate $r_i^{\text{all}} = r_i + C_i$ to the ith UE in β.

Algorithm 1 The ith SUE $\in \mathcal{Q}_s$ algorithm

---

loop

Send application utility parameters k_i, a_i, b_i, $r_i^{\max}$, and u_i^{req} to the SUE's in band small cell's eNodeB.

Receive the final allocated rate $r_i^{s,\text{all}}$ from the small cell s eNodeB or from the macro cell's eNodeB.

end loop

Algorithm 2 The ith MUE $\in \mu$ algorithm

loop
  Send application utility parameters k_i, a_i, b_i, and $r_i^{\max}$ to the macro cell's eNodeB.
  Receive the final allocated rate r_i^{all} from the macro cell's eNodeB.
end loop

Algorithm 3 Small cell s eNodeB algorithm

loop
  Initialize $\mathcal{Q}_{sB} = \emptyset$; $r_i^{\text{all}} = 0$.
  Receive application utility parameters k_i, a_i, b_i, $r_i^{\max}$, and u_i^{req} from all SUEs in $\mathcal{Q}_s$.
  Solve $\mathbf{r}^s = \arg\max_{\mathbf{r}^s} \sum_{i=1}^{|\mathcal{Q}_s|} \log U_i(r_i^s) - p^s(\sum_{i=1}^{|\mathcal{Q}_s|}(r_i^s) - R_s)$.
  Let $r_i^{s,\text{all}} = r_i^s$ be the rate allocated by the s small cell's eNodeB to each user in $\mathcal{Q}_s$.
  Calculate the SUE utility $U_i(r_i^{s,\text{all}})$ $\forall i \in \mathcal{Q}_s$
  for SUE $i \leftarrow 1$ to $|\mathcal{Q}_s|$ **do**
    if $U_i(r_i^{s,\text{all}}) < u_i^{\text{req}}$ **then**
      $\mathcal{Q}_{sB} = \mathcal{Q}_{sB} \bigcup \text{SUE}\{i\}$
      Send SUE i parameters k_i, a_i, b_i, $r_i^{\max}$, and $r_i^{s,\text{all}}$ to the macro cell's eNodeB
    else
      Allocate rate $r_i^{\text{all}} = r_i^{s,\text{all}}$ to SUE i
    end if
  end for
end loop

Algorithm 4 Macro cell's eNodeB algorithm

loop
  Initialize $C_i = 0$; $r_i^{\text{all}} = 0$.
  for $s \leftarrow 1$ to S **do**
    Receive application utility parameters k_i, a_i, b_i, $r_i^{\max}$, and $r_i^{s,\text{all}}$ for all SUEs in $\mathcal{Q}_{sB}$ from small cell s eNodeB.
    $C_i = r_i^{s,\text{all}}$ $\forall i \in \mathcal{Q}_{sB}$
  end for
  Create user group $\mathcal{Q} = \bigcup_{s=1}^{S} \mathcal{Q}_{sB}$
  Create user group $\beta = \mu \bigcup \mathcal{Q}$
  Solve $\mathbf{r} = \arg\max_{\mathbf{r}} \sum_{i=1}^{|\beta|} \log U_i(r_i + C_i) - p^B(\sum_{i=1}^{|\beta|}(r_i) - R_B)$.
  Allocate $r_i^{\text{all}} = r_i + C_i$ to each UE i in β
end loop

5.4 Spectrum Sharing Simulation Results

Algorithms 1–4 were applied in C++ (5.6) to multiple utility functions with different parameters. Simulation results showed convergence to the global optimal rates. In this section, we consider a macro cell with one eNodeB. Within the coverage area of the macro cell, there exists one small cell s. Four SUEs are located under the coverage area of the small cell s with UEs indexes $\{1, 2, 3, 4\}$. The SUEs user group is given by $\mathcal{Q}_s = \{1, 2, 3, 4\}$. Four MUEs are located under the coverage area of the macro cell's eNodeB but not within the small cell. The MUEs user group is

Table 5.1 Users and their applications utilities

User's index	User's type	Applications utilities parameters
UE1 $i = \{1\}$	SUE	Sig2 $a_i = 3,\ b_i = 20,\ u_i^{req} = 0.8$
UE2 $i = \{2\}$	SUE	Sig3 $a_i = 1,\ b_i = 30,\ u_i^{req} = 0.8$
UE3 $i = \{3\}$	SUE	Log2 $k_i = 3,\ r_i^{max} = 100,\ u_i^{req} = 0.5$
UE4 $i = \{4\}$	SUE	Log3 $k_i = 0.5,\ r_i^{max} = 100,\ u_i^{req} = 0.5$
UE5 $i = \{5\}$	MUE	Sig1 $a_i = 5,\ b_i = 10$
UE6 $i = \{6\}$	MUE	Sig3 $a_i = 1,\ b_i = 30$
UE7 $i = \{7\}$	MUE	Log1 $k_i = 15,\ r_i^{max} = 100$
UE8 $i = \{8\}$	MUE	Log3 $k_i = 0.5,\ r_i^{max} = 100$

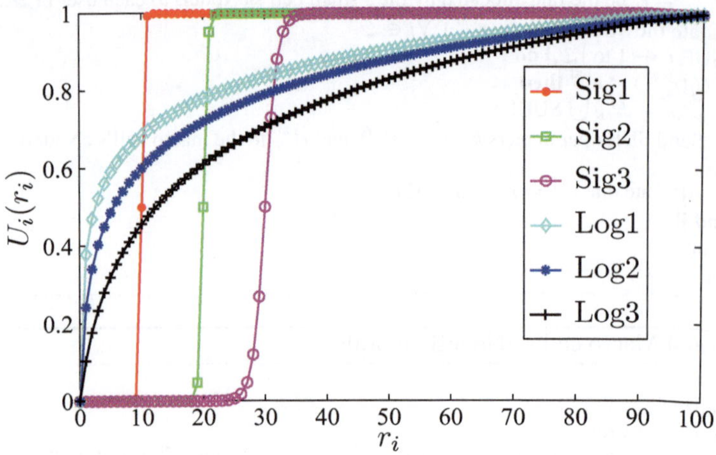

Fig. 5.2 The users utility functions $U_i(r_i)$ used in the simulation (three sigmoidal-like functions and three logarithmic functions)

given by $\mu = \{5, 6, 7, 8\}$. Each UE whether it is SUE or MUE is running either real-time application or delay-tolerant application. Each of the SUEs' applications utilities has a minimum required utility that is given by u_i^{req} that is equivalent to the C_i value for that user whereas MUEs do not have minimum required utilities for their applications. The UEs' indexes, types, and applications utilities parameters are listed in Table 5.1. Figure 5.2 shows the sigmoidal-like utility functions and the logarithmic utility functions used to represent the SUEs and MUEs applications.

5.4.1 Small Cell Allocated Rates and Users QoE

In the following simulations, the small cell's carrier total rate R_s takes values between 10 and 100 with step of 10. In Fig. 5.3, we show the small cell's allocated

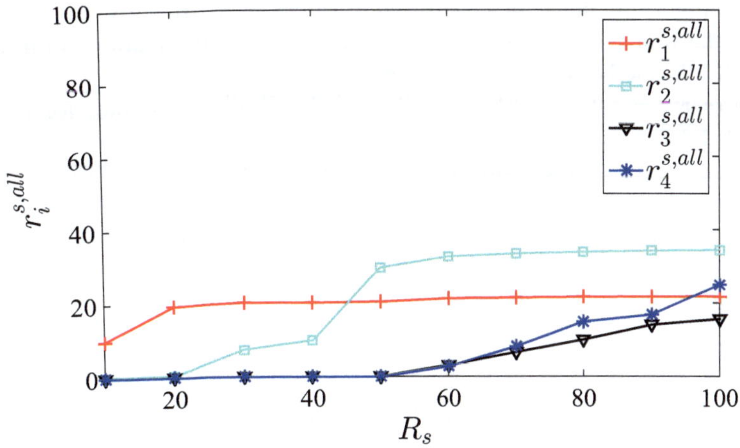

(a)

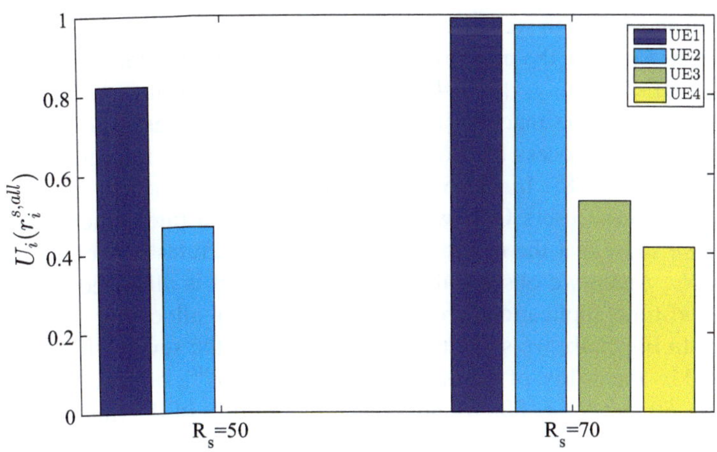

(b)

Fig. 5.3 The small cell's eNodeB allocated rates with $10 < R_s < 100$ and users' QoE when $R_s = 50$ and $R_s = 70$. (**a**) The rates $r_i^{s,\text{all}}$ allocated by the small cell's eNodeB to users in $\mathcal{Q}_s$ with $10 < R_s < 100$. (**b**) Users' QoE represented by the utility of user's application of its allocated rate $U_i(r_i^{s,\text{all}})$ when $R_s = 50$ and $R_s = 70$

rates $r_i^{s,\text{all}}$ for users in $\mathcal{Q}_s$ with different values of the small cell's carrier total rate R_s and the users QoE with the small cell allocated rates when $R_s = 50$ and $R_s = 70$. In Fig. 5.3a, we show that users running real-time applications are given priority when allocating the small cell's resources due to their sigmoidal-like utility function nature. We also observe that none of the UEs is allocated zero resources because we used a utility proportional fairness approach. We also show how the proposed rate allocation algorithm converges for different values of R_s. In Fig. 5.3b, we show the QoE for the four SUEs which is represented by their applications utilities of the small cell allocated rates $U_i(r_i^{s,\text{all}})$ when $R_s = 50$ and $R_s = 70$. We notice that in the case of $R_s = 50$, the utilities of the small cell allocated rates for UE2, UE3, and UE4 did not reach the minimum required utilities for these SUEs whereas in the case of $R_s = 70$ the utility of the small cell allocated rate for UE4 did not reach the minimum required utility for that SUE. Therefore, based on the proposed algorithm the network operator will request additional resources for these UEs from the macro cell's eNodeB and these UEs will be allocated additional resources based on a resource allocation with carrier aggregation scenario.

5.4.2 *Macro Cell Allocated Rates and Users QoE*

In the following simulations, the macro cell's carrier total rate R_B takes values between 10 and 100 with step of 10 and R_s is fixed at 50. As discussed in Sect. 5.4.1, in the case of $R_s = 50$ the network operator requests additional resources for three SUEs (i.e., UEs in $\mathcal{Q}_{sB} = \{2, 3, 4\}$) as they did not reach their minimum required utilities. Therefore, the macro cell's eNodeB performs a resource allocation with carrier aggregation process to allocate resources to the UEs in user group β where $\beta = \{2, 3, 4, 5, 6, 7, 8\}$. In Fig. 5.4, we show the final allocated rates r_i^{all} for the UEs in β and these users QoE with the final allocated rates when $R_B = 80$. In Fig. 5.4a, we show that the macro cell's final allocated rates converge for different values of R_B. Again we observe that none of the users is allocated zero resources and that real-time applications are given priority when allocating the macro cell's resources. In Fig. 5.4b, we show the QoE for each of the seven UEs in β which is represented by the utility, of the final allocated rate $U_i(r_i^{\text{all}})$, of the user's application when $R_s = 50$ and $R_B = 80$. We notice that the utilities of the final allocated rates for the three SUEs in $\mathcal{Q}_{sB}$ (i.e., UE {2,3,4}) exceed the minimum required utilities for these SUEs because of the additional resources allocated to these users by the macro cell's eNodeB.

5.5 Summary and Conclusion

In this chapter, we proposed a spectrum sharing approach for sharing the Federal under-utilized 3.5 GHz spectrum with commercial users and used the multi-stage resource allocation with CA algorithms to allocate the macro cell and small cells resources optimally among users under their coverage area. Users located under the coverage area of the small cells are allocated resources by the small cells' eNodeBs whereas both the macro cell users and the small cells' users that did not reach their minimum required utilities, by their small cells' allocated rates, are allocated resources by the macro cell's eNodeB based on carrier aggregation. We showed

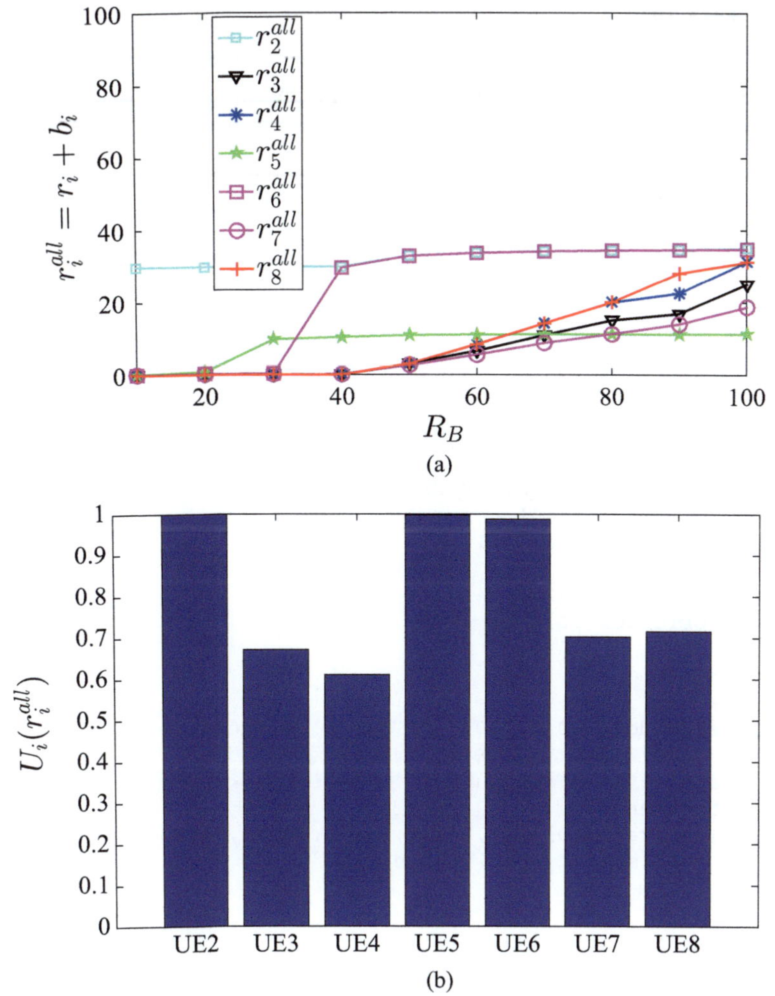

Fig. 5.4 The total aggregated rates $r_i^{\text{all}} = r_i + C_i$ allocated by the macro cell's eNodeB to users in β with $10 < R_B < 100$ when $R_s = 50$ and the users' QoE when $R_B = 80$ and $R_s = 50$. (**a**) The aggregated rates $r_i^{\text{all}} = r_i + C_i$ allocated by the macro cell's eNodeB to users in β when $R_s = 50$. (**b**) Users' QoE represented by the utility of user's application of its allocated rate $U_i(r_i^{\text{all}})$ when $R_B = 80$ and $R_s = 50$

through simulations that the proposed algorithm converges to the optimal rates. We also showed that small cells' users can achieve their minimum required QoE by using the proposed spectrum sharing approach.

5.6 C++ Code

Some functions in this code are listed in the Appendix.

```
#include<iostream>
#include<math.h>
#include <string> //This is for string
#include <windows.h> //This is for the console code
using namespace std;

float c_i[3];
const float lamda = 0.0000001;
float low = 0.01;
float rate_optimal_all_R[4][10];
float rate_optimal_all_R_B[7][10];
int counter = 0;
float rate_optimal_all_R_C1[4][10];
float Utility[4][10];
float UtilityMacro[7][10];
float price1[10];
float priceMacro[10];
float price1C1[10];

//main
int main(){
    SetWindow(100,10000); //This sets the size of the window to Width 100, Height 1000
    cout <<"prog is running...Done! \n"<<endl;
    cout<<endl;
    float rate[10]={10,20,30,40,50,60,70,80,90,100}; //eNodeB's maximum achievable rate R
    int time=0;

    for (int i=0; i<4; i++){
        cout<<endl;
        cout<<"user # "<<i<<endl;
        cout<<endl;
        for(int j=0;j<10;j++){
        float rate1=rate[j];
        int time=j;
    rateCalc1(time,rate1);
        }

    cout<<endl;
         cout<<" optimal rate ri for user # "<<i<<endl;
         for(int n = 0; n<10; n++){

        cout<<rate_optimal_all_R[i][n]<<" ";
         }
    }

        cout<<endl;
```

```
        cout<<" Carrier1 prices "<<endl;
        for(int n = 0; n<10; n++){  // small cell's eNodeB
            prices for different values of R
       cout<<price1[n]<<" ";
        }
    cout<<endl;
    cout<<endl;
        cout<<" Utility of Small Cell Users "<<endl;
         for(int i = 0; i<4; i++){
             cout<<endl;
        cout<<" utility for user # "<<i<<endl;
        for(int n = 0; n<10; n++){
       cout<<Utility[i][n]<<" ";
        }
         }
    cout<<endl;
    cout<<endl;
    //////////////////////////////////
    int time2 = 0;
    for (int i=0; i<7; i++){  //Macro cell's 8 users
        cout<<endl;
        cout<<"user # "<<i<<endl;
        cout<<endl;
        for(int j=0;j<10;j++){ //Macro cell's maximum
            achievable rate R
        float rate2=rate[j];
        int time2=j;
    rateCalc2(time2,rate2);
        }

    cout<<endl;
        cout<<" optimal rate ri for user # "<<i<<endl;
         for(int n = 0; n<10; n++){
        cout<<rate_optimal_all_R_B[i][n]<<" ";
         }
    }

        cout<<endl;
         cout<<"Macro Cell Carrier prices "<<endl; //Macro
            cell's eNodeB prices for different values of R
         for(int n = 0; n<10; n++){
        cout<<priceMacro[n]<<" ";
         }

    cout<<endl;
    cout<<endl;
        cout<<" Utility of Macro Cell Users "<<endl;
         for(int i = 0; i<7; i++){
             cout<<endl;
        cout<<" utility for user # "<<i<<endl;
        for(int n = 0; n<10; n++){
       cout<<UtilityMacro[i][n]<<" "; //Macro cell users'
            utilities of their allocated rates
         }
```

```
        }

    cout<<endl;
    cout<<endl;

     system("PAUSE");
    return 0;
}

//---------------------------------------------------------------//

//For Small Cell Resource Allocation//
//This function is called by main
void rateCalc1(int time, float rate1){
    float weight[4] ={1,1,1,1}; //Initialize small cell users' bids
    float prev_Weight[4] = {0,0,0,0};
    float p_n;
    float p_prev = 0;
    float k_i[4] = {3,.5,15,.5};
    float a_i[4] = {3,1,5,1};
    float b_i[4] = {20,30,10,30};
    float opt1[4] = {0,0,0,0};
    float u_req[4] = {0,30,0,15}; //Users' minimum required utilities
    float opt2[4]={0,0,0,0};
    float rateArray[4]={0,0,0,0};
    float R=rate1;
    float weight2[4]={0,0,0,0};
    float r_i_optimal[4];
    int N_users=4;
    int counter=0;

    p_n = shadowPrice(weight,R,N_users);

    while (((abs(weight[0] - prev_Weight[0]) ≥ lamda) || (abs(weight[1] - prev_Weight[1]) ≥ lamda) || (abs(weight[2] - prev_Weight[2]) ≥ lamda) || (abs(weight[3] - prev_Weight[3]) ≥ lamda))){

        //Finding ri
        for(int i = 0; i <4; i++){
            r_i_optimal[i] = weight[i]/p_n;
        }
        //Finding ri_optimal
        for(int j = 0; j<2; j++){
            rateArray[j] = get_roots1(a_i[j],b_i[j],p_n,opt1[j],opt2[j],R);
        }

        for(int j = 2; j<4; j++){
            rateArray[j] = get_roots2(k_i[j-2],p_n,opt1[j],opt2[j],R);
        }
```

```

        for(int k = 0; k<4; k++){
            prev_Weight[k] = weight[k];
        }

        /////////Calculate users' bids/////////
        for(int k = 0; k<4; k++){

            weight2[k] = p_n * rateArray[k];
        }
        /////////Fluctuation Decay Condition/////////
        for(int k = 0; k<4; k++){
            float exp = 2.718281828459;

            float Del_W= 5*pow(exp, (-counter/10));
            if (abs(weight2[k]-prev_Weight[k])>Del_W){ // Check fluctuation condition
            weight[k] = prev_Weight[k]+(((weight2[k] - prev_Weight[k])/abs(weight2[k]-prev_Weight[k]))*Del_W); //Calculate new bid

            }
            else
            weight[k]=weight2[k];

        }

        p_prev = p_n;
        p_n = shadowPrice(weight,R,N_users);
        //cout<<p_n<<"--p_n_2"<<endl;
        counter++;
    }

    for(int i=0;i<4;i++){
        rate_optimal_all_R[i][time]=rateArray[i];
        rate_optimal_all_R_C1[i][time]=rateArray[i];
    }
    price1[time]=p_n;
    price1C1[time]=p_n;
    for(int j = 0; j<2; j++){
        Utility[j][time] = U_Sig(rateArray[j],a_i[j],b_i[j]);
    }
    for(int j = 2; j<4; j++){
        Utility[j][time] = U_Log(rateArray[j],k_i[j-2]);
    }
}

//------------------------------------------------------------//
//For Macro Cell Resource Allocation with Carrier Aggregation //
//This function is called by main
void rateCalc2(int time, float rate1){
    float weight[7] ={1,1,1,1,1,1,1}; //Initial bids for Macro cell users
```

```
    float prev_Weight[7] = {0,0,0,0,0,0,0}; //Initial
        previous bids
    float p_n;  //Price per unit resource
    float p_prev = 0;
    float k_i[4] = {3,.5,15,.5}; //For logarithmic utility
        parameters
    float a_i[3] = {1,5,1};     //For Sigmoidal-like utility
        parameters
    float b_i[3] = {30,10,30}; //For Sigmoidal-like utility
        parameters
    float opt1[7] = {29.871,0.001,0.001,0,0,0,0}; //Rates
        allocated to small cell users by small cell/ eNodeB
    float opt2[7] = {0,0,0,0,0,0,0};
    float rateArray[7] = {0,0,0,0,0,0,0};
    float R_B=rate1;
    float weight2[7] = {0,0,0,0,0,0,0};
    float r_i_optimal[7];
    int N_users = 7;
    int counter = 0;

    p_n = shadowPrice(weight,R_B,N_users);

    while (((abs(weight[0] - prev_Weight[0]) ≥ lamda) || (abs
        (weight[1] - prev_Weight[1]) ≥ lamda) || (abs(weight
        [2] - prev_Weight[2]) ≥ lamda) || (abs(weight[3] -
        prev_Weight[3]) ≥ lamda) || (abs(weight[4] -
        prev_Weight[4]) ≥ lamda) || (abs(weight[5] -
        prev_Weight[5]) ≥ lamda) || (abs(weight[6] -
        prev_Weight[6]) ≥ lamda))){
        //Finding ri
        for(int i = 0; i <7; i++){
            r_i_optimal[i] = weight[i]/p_n;
        }

        //Finding ri_optimal
        for(int j = 0; j<1; j++){
            rateArray[j] = get_roots1(a_i[j],b_i[j],p_n,opt1[
                j],opt2[j],R_B);
        }

        for(int j = 1; j<3; j++){
            rateArray[j] = get_roots2(k_i[j-1],p_n,opt1[j],
                opt2[j],R_B);
        }

        for(int j = 3; j<5; j++){
            rateArray[j] = get_roots1(a_i[j-2],b_i[j-2],p_n,
                opt1[j],opt2[j],R_B);
        }

        for(int j = 5; j<7; j++){
            rateArray[j] = get_roots2(k_i[j-3],p_n,opt1[j],
                opt2[j],R_B);
        }
```

```
        for(int k = 0; k<7; k++){

            prev_Weight[k] = weight[k];
        }
        /////////Calculate users' bids/////////
        for(int k = 0; k<7; k++){

            weight2[k] = p_n * rateArray[k];
        }

        /////////Fluctuation Decay Condition/////////
        for(int k = 0; k<7; k++){
            float exp = 2.718281828459;

            float Del_W= 5*pow(exp, (-counter/10));
            if (abs(weight2[k]-prev_Weight[k])>Del_W){ //
                Check fluctuation condition
            weight[k] = prev_Weight[k]+(((weight2[k] -
                prev_Weight[k])/abs(weight2[k]-prev_Weight[k])
                )*Del_W); //Calculate new bid

            }
            else
                weight[k]=weight2[k];
        }
        p_prev = p_n;
        p_n = shadowPrice(weight,R_B,N_users);
        counter++;
    }

    for(int i=0;i<7;i++){
        rate_optimal_all_R_B[i][time]=rateArray[i]+opt1[i];
    }
    price1[time]=p_n;
    price1C1[time]=p_n;
        for(int j = 0; j<1; j++){
            UtilityMacro[j][time] = U_Sig(rateArray[j]+opt1[j
                ],a_i[j],b_i[j]);
        }
        for(int j = 1; j<3; j++){
            UtilityMacro[j][time] = U_Log(rateArray[j]+opt1[j
                ],k_i[j-1]);
        }
        for(int j = 3; j<5; j++){
            UtilityMacro[j][time] = U_Sig(rateArray[j]+opt1[j
                ],a_i[j-2],b_i[j-2]);
        }
        for(int j = 5; j<7; j++){
            UtilityMacro[j][time] = U_Log(rateArray[j]+opt1[j
                ],k_i[j-3]);
        }
}
```

References

1. Executive Office of the President, President's Council of Advisors on Science and Technology (PCAST), *Realizing the Full Potential of Government-Held Spectrum to Spur Economic Growth* (PCAST, Washington, DC, 2012)
2. M. Richards, J. Scheer, W. Holm, *Principles of Modern Radar* (Scitech, New York, 2010)
3. Federal Communications Commission (FCC), In the matter of revision of parts 2 and 15 of the commission's rules to permit unlicensed national information infrastructure (U-NII) devices in the 5 GHz band. MO and O, ET Docket No. 03–122, June 2006
4. Federal Communications Commission, Proposal to create a Citizen's broadband service in the 3550–3650 MHz band. Report (2012)
5. Federal Communications Commission (FCC), Connecting America: the national broadband plan (Online, 2010)
6. G. Yuan, X. Zhang, W. Wang, Y. Yang, Carrier aggregation for LTE-advanced mobile communication systems. IEEE Commun. Mag. **48**, 88–93 (2010)
7. Z. Shen, A. Papasakellariou, J. Montojo, D. Gerstenberger, F. Xu, Overview of 3GPP LTE-advanced carrier aggregation for 4G wireless communications. IEEE Commun. Mag. **50**, 122–130 (2012)
8. *Frequency Spectrum Wall Chart* (Commerce Dept., National Telecommunications and Information Administration, Office of Spectrum Management, 2016)
9. National Telecommunications and Information Administration (NTIA), An assessment of the near-term viability of accommodating wireless broadband systems in the 1675–1710 MHz, 1755–1780 MHz, 3500–3650 MHz, 4200–4220 MHz, and 4380–4400 MHz bands (Fast Track Report), Online, October 2010
10. NTIA, An assessment of the near-term viability of accommodating wireless broadband systems in the 1675–1710 MHz, 1755–1780 MHz, 3500–3650 MHz, 4200–4220 MHz and 4380–4400 MHz bands. Report (2010)
11. National Telecommunications and Information Administration (NTIA), Analysis and resolution of RF interference to radars operating in the band 2700–2900 MHz from broadband communication transmitters, Online, October 2012
12. M. Cotton, R. Dalke, Spectrum occupancy measurements of the 3550–3650 MHz maritime radar band near San Diego, California (2014)
13. H. Shajaiah, A. Abdelhadi, C. Clancy, An application-aware spectrum sharing approach for commercial use of 3.5 GHz spectrum, in *2016 Annual IEEE Systems Conference (SysCon)* (2016), pp. 1–7
14. G. Tychogiorgos, A. Gkelias, K.K. Leung, Utility-proportional fairness in wireless networks, in *International Symposium on Personal, Indoor, and Mobile Radio Communications* (IEEE, New York, 2012), pp. 839–844
15. A. Abdelhadi, C. Clancy, A utility proportional fairness approach for resource allocation in 4G-LTE, in *IEEE International Conference on Computing, Networking, and Communications (ICNC), CNC Workshop* (2014)
16. A. Abdelhadi, C. Clancy, A robust optimal rate allocation algorithm and pricing policy for hybrid traffic in 4G-LTE, in *IEEE International Symposium on Personal, Indoor, and Mobile Radio Communications (PIMRC)* (2013)
17. Y. Wang, A. Abdelhadi, A QoS-based power allocation for cellular users with different modulations, in *2016 International Conference on Computing, Networking and Communications (ICNC)* (2016), pp. 1–5
18. A. Abdelhadi, H. Shajaiah, Optimal resource allocation for smart phones with multiple applications with MATLAB instructions. Technical Report (2016)
19. A. Abdelhadi, H. Shajaiah, Application-aware resource allocation with carrier aggregation using MATLAB. Technical Report (2016)

20. Z. Kbah, A. Abdelhadi, Resource allocation in cellular systems for applications with random parameters, in *2016 International Conference on Computing, Networking and Communications (ICNC)* (2016), pp. 1–5
21. A. Abdelhadi, H. Shajaiah, Optimal resource allocation for cellular networks with MATLAB instructions. CoRR, abs/1612.07862 (2016)
22. F. Wilson, I. Wakeman, W. Smith, Quality of service parameters for commercial application of video telephony, in *Proceedings of the Human Factors in Telecommunications Symposium* (Darmstadt, 1993), pp. 139–148
23. S. Shenker, Fundamental design issues for the future internet. IEEE J. Sel. Areas Commun. **13**(7), 1176–1188 (1995)
24. Y. Wang, A. Abdelhadi, T. C. Clancy, Optimal power allocation for LTE users with different modulations, in *2016 Annual IEEE Systems Conference (SysCon)* (2016), pp. 1–5
25. M. Ghorbanzadeh, A. Abdelhadi, C. Clancy, Distributed resource allocation, in *Cellular Communications Systems in Congested Environments* (Springer, Berlin, 2017), pp. 61–91
26. J.-W. Lee, R.R. Mazumdar, N.B. Shroff, Downlink power allocation for multi-class wireless systems. IEEE/ACM Trans. Netw. **13**, 854–867 (2005)
27. A. Kumar, A. Abdelhadi, T.C. Clancy, A delay efficient multiclass packet scheduler for heterogeneous M2M uplink, in *IEEE MILCOM* (2016)
28. A. Kumar, A. Abdelhadi, T.C. Clancy, An online delay efficient packet scheduler for M2M traffic in industrial automation, in *IEEE Systems Conference* (2016)
29. A. Kumar, A. Abdelhadi, T.C. Clancy, A delay optimal MAC and packet scheduler for heterogeneous M2M uplink. CoRR, abs/1606.06692 (2016)
30. A. Kumar, A. Abdelhadi, T.C. Clancy, A delay-optimal packet scheduler for M2M uplink, in *IEEE MILCOM* (2016)

Chapter 6
RA with CA for a Cellular System Sharing Spectrum with S-Band Radar

As a result of the high demand for spectrum by commercial wireless operators, federal agencies are now willing to share their spectrum with commercial users [1–3]. The 3550–3650 MHz band, currently used for military radar operations, is identified for spectrum sharing between military radars and communication systems, according to the NTIA's 2010 Fast Track Report [4–7] and the recommendations of PCAST [8] and FCC [9–12]. This band is very favorable for commercial cellular systems such as LTE-Advanced systems. However, radar interference to cellular systems is a cause of concern for commercial operators and thus innovative methods are required to make spectrum sharing between radars and cellular systems a reality.

In this chapter, we consider an LTE-Advanced cellular system sharing the 3550–3650 MHz band with a MIMO radar [13]. The LTE-Advanced cellular system has N_{BS} base stations. In order to mitigate radar interference, a spectrum sharing algorithm is proposed. The algorithm selects the best interference channel for radar's signal projection to mitigate radar interference to the ith BS. We consider a MIMO collocated radar mounted on a ship. Collocated radars have improved spatial resolution over widely spaced radars [14]. The LTE cellular system operates in its regular licensed band and shares the 3.5 GHz band with a MIMO radar in order to increase its capacity such that the two systems do not cause interference to each other. We focus on finding an optimal solution for the resource allocation with carrier aggregation problem to allocate the LTE-Advanced BS/eNodeB and the available MIMO radar resources optimally among users subscribing for a service in the cellular cell coverage area. Each user is assigned a utility function based on the application running on its UE. Real-time applications are represented by sigmoidal-like utility functions whereas delay-tolerant applications are represented by logarithmic utility functions. Real-time applications are given the priority when

The content in this chapter is reproduced with permission after modifications (License numbers 4078230325782 and 4078221507027). For the original article please refer to [15, 16].

H. Shajaiah et al., *Resource Allocation with Carrier Aggregation in Cellular Networks*, DOI 10.1007/978-3-319-60540-1_6

allocating resources. A resource allocation with carrier aggregation algorithm is proposed in this chapter to allocate the LTE-Advanced eNodeB and the MIMO radar resources optimally among users. The proposed algorithm is performed in two stages, the LTE-Advanced eNodeB resources are first allocated to users subscribing for a service and then the available MIMO radar resources are allocated to the same users. The algorithm employs a proportional fairness approach in its two stages to guarantee that no user is allocated zero resources and gets dropped. The contributions in this chapter are summarized as:

- We present a spectrum sharing scenario between a MIMO radar and LTE system with multiple base stations and propose a channel-selection algorithm to select the best channel for radar's signal projection that maintains a minimum degradation in the radar performance while causing no interference to the LTE BS. We also present our null-space projection (NSP) algorithm that performs the null-space computation.
- We present a resource allocation optimization problem with carrier aggregation to allocate the LTE-Advanced and the MIMO radar carriers resources optimally among users running real-time or delay-tolerant applications.
- We propose a two-stage resource allocation algorithm to allocate the two carriers resources optimally among users. First, the LTE-Advanced eNodeB and the UEs collaborate to allocate an optimal rate to each UE. Once the LTE-Advanced eNodeB finishes allocating resources to the UEs, the eNodeB then allocates the MIMO radar's available resources to these UEs.

The remainder of this chapter is organized as follows. Section 6.1 discusses the spectrum sharing scenario between MIMO radar and LTE cellular system. In Sect. 6.2, we describe collocated MIMO radars. In Sect. 6.3, we present our channel-selection and NSP algorithms and explain the projection of radar signal onto the null space of the selected interference channel. In Sect. 6.4 we present our resource allocation with carrier aggregation optimization problem using a utility proportional fairness approach. In Sect. 6.5 we present our two-stage distributed robust resource allocation with carrier aggregation algorithm for the optimization problem. Section 6.6 discusses simulation setup and provides quantitative results along with discussion. Section 6.7 concludes the paper.

6.1 System Model

We consider a collocated MIMO radar and a MIMO LTE communication system. The two systems are the primary users of the 3550–3650 MHz band under consideration. The MIMO radar has M_T transmit antennas and M_R receive antennas. The LTE communication system has N_{BS} base stations, each BS is equipped with N_T^{BS} transmit antennas and N_R^{BS} receive antennas, with the ith BS supporting K_i^{UE} user equipments (UE)s. Each UE is equipped with N_T^{UE} transmit antennas and N_R^{UE} receive antennas. The collocated radars give better target parameter identifiability and improved spatial resolution as their antenna spacing is on the order of half the wavelength of the carrier [14]. The MIMO radar projects its signal onto the null space of the interference channel while illuminating a target. The MIMO radar is sharing N_{BS} interference channels $\mathbf{H}_i^{N_R^{\text{BS}} \times M_T}$ with the LTE system. Let $\mathbf{x}_{\text{Radar}}(t)$ and $\mathbf{x}_j^{\text{UE}}(t)$ be the signals transmitted from the MIMO radar and the jth UE in the ith cell, respectively. The received signal at the ith BS receiver can be written as

$$\mathbf{y}_i(t) = \mathbf{H}_i^{N_R^{\text{BS}} \times M_T} \mathbf{x}_{\text{Radar}}(t) + \sum_j \mathbf{H}_j^{N_R^{\text{BS}} \times N_T^{\text{UE}}} \mathbf{x}_j^{\text{UE}}(t) + \mathbf{w}(t)$$

$$\text{for } 1 \leq i \leq N_{\text{BS}} \text{ and } 1 \leq j \leq K_i^{\text{UE}}$$

where $\mathbf{w}(t)$ is the additive white Gaussian noise. In order to avoid interference to the ith LTE BS, the MIMO radar maps $\mathbf{x}_{\text{Radar}}(t)$ onto the null space of $\mathbf{H}_i^{N_R^{\text{BS}} \times M_T}$. Figure 6.1 shows a spectrum sharing scenario between a maritime MIMO radar and an LTE cellular system where the MIMO radar is sharing N_{BS} interference channels $\mathbf{H}_i^{N_R^{\text{BS}} \times M_T}$ with the LTE system.

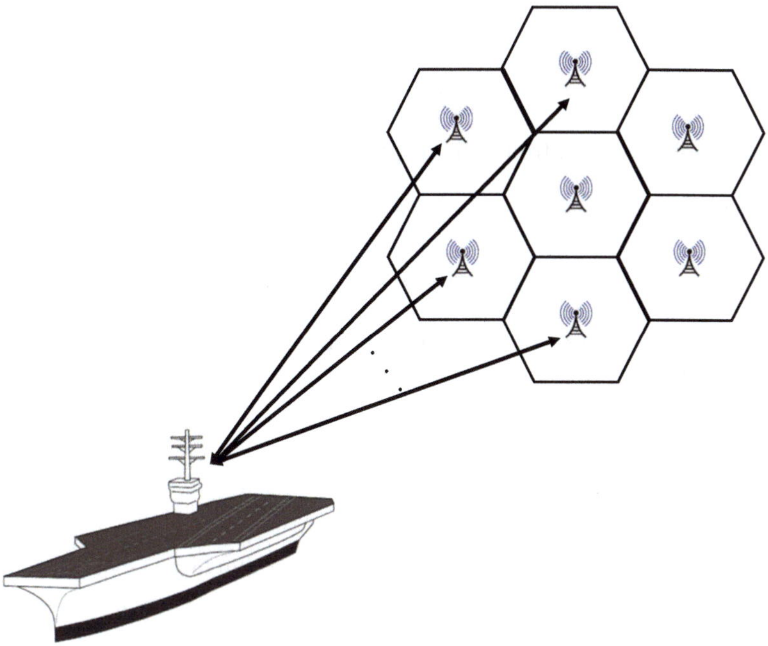

Fig. 6.1 Spectrum sharing scenario between LTE cellular system and a maritime MIMO radar

6.2 Radar-LTE Spectrum Sharing Approach

The MIMO radar we consider is a collocated MIMO radar with M_T transmit antennas and M_R receive antennas. Let $\mathbf{x}_{\text{Radar}}(t)$ be the signal transmitted from the MIMO radar, defined as

$$\mathbf{x}_{\text{Radar}}(t) = \left[x_1(t)e^{j\omega_c t}\ x_2(t)e^{j\omega_c t}\ \cdots\ x_{M_T}e^{j\omega_c t}(t)\right]^T$$

where ω_c is the carrier angular frequency, $x_k(t)$ is the baseband signal from the kth transmit element, and $t \in [0, T_o]$ with T_o being the observation time. The radar transmit steering vector is defined as

$$\mathbf{a}_T(\theta) \triangleq \left[e^{-j\omega_c\tau_{T_1}(\theta)}\ e^{-j\omega_c\tau_{T_2}(\theta)}\ \ldots\ e^{-j\omega_c\tau_{T_{M_T}}(\theta)}\right]^T$$

the radar receive steering vector is defined as

$$\mathbf{a}_R(\theta) \triangleq \left[e^{-j\omega_c\tau_{R_1}(\theta)}\ e^{-j\omega_c\tau_{R_2}(\theta)}\ \ldots\ e^{-j\omega_c\tau_{R_{M_R}}(\theta)}\right]^T$$

and the transmit-receive steering matrix is defined as

$$\mathbf{A}(\theta) \triangleq \mathbf{a}_R(\theta)\mathbf{a}_T^T(\theta).$$

Then, the signal received from a single point target at an angle θ is given by

$$\mathbf{y}_{\text{Radar}}(t) = \alpha\, e^{-j\omega_D t}\, \mathbf{A}(\theta)\, \mathbf{x}_{\text{Radar}}(t - \tau(t))$$

where $\tau(t) = \tau_r = \tau_{T_k}(t) + \tau_{R_l}(t)$ is the sum of propagation delays between the target and the kth transmit element and between the target and the lth receive element, respectively; and α represents the complex path loss including the propagation loss and the coefficient of reflection.

6.3 Spectrum Sharing Algorithms

In this section, we present a channel-selection algorithm to select the best interference channel on which radar signals are projected. We also present NSP algorithm that performs the null-space computation.

6.3.1 Channel-Selection Algorithm

Our channel-selection algorithm, shown in Algorithm 1, selects the best interference channel onto which radar signals are projected. Based on our system model, we assume that there exist N_{BS} interference channels $\mathbf{H}_i$, $i = 1, 2, \ldots, N_{\text{BS}}$ between the MIMO radar and the LTE system. Our goal is to select the best interference channel defined as

$$i_{\min} \triangleq \underset{1 \le i \le N_{\text{BS}}}{\arg\min} \|\mathbf{x}_{\text{Radar}} - \mathbf{P}_{\mathbf{V}_i}\mathbf{x}_{\text{Radar}}\|$$

$$\mathbf{H}_{\text{Best}} \triangleq \mathbf{H}_{i_{\min}}$$

we also seek to avoid the worst interference channel defined as

$$i_{\max} \triangleq \underset{1 \le i \le N_{\text{BS}}}{\arg\max} \|\mathbf{x}_{\text{Radar}} - \mathbf{P}_{\mathbf{V}_i}\mathbf{x}_{\text{Radar}}\|$$

$$\mathbf{H}_{\text{Worst}} \triangleq \mathbf{H}_{i_{\max}}$$

where $(\mathbf{x}_{\text{Radar}} - \mathbf{P}_{\mathbf{V}_i}\mathbf{x}_{\text{Radar}})$ is the difference between the original radar waveform $\mathbf{x}_{\text{Radar}}$ and the radar waveform projected onto the null space of $\mathbf{H}_i$ and the Euclidean norm of $(\mathbf{x}_{\text{Radar}} - \mathbf{P}_{\mathbf{V}_i}\mathbf{x}_{\text{Radar}})$ is defined as

$$\begin{aligned}
&\|\mathbf{x}_{\text{Radar}} - \mathbf{P}_{\mathbf{V}_i}\mathbf{x}_{\text{Radar}}\| \\
&\quad = \sqrt{(\mathbf{x}_{\text{Radar}} - \mathbf{P}_{\mathbf{V}_i}\mathbf{x}_{\text{Radar}})^H(\mathbf{x}_{\text{Radar}} - \mathbf{P}_{\mathbf{V}_i}\mathbf{x}_{\text{Radar}})}.
\end{aligned}$$

We use the blind null-space learning algorithm introduced in [17] to estimate the channel state information (CSI) of the N_{BS} interference channels at the MIMO radar. The projection matrix $\mathbf{P}_{\mathbf{V}_i}$ of each of the N_{BS} interference channels is then found using Algorithm 2. Once Algorithm 1 receives the projection matrices of the interference channels, it selects the best interference channel $\check{\mathbf{H}}$ and sends it to Algorithm 2 for NSP of radar signals. Selecting the best interference channel using our channel-selection algorithm (i.e., Algorithm 1) guarantees minimum degradation in the performance of the radar while maintaining no interference to the LTE BS.

6.3.2 Null-Space Projection (NSP) Algorithm

In this section, we present our proposed null-space projection algorithm. We also explain the projection of radar signals onto null space of the best interference channel selected using Algorithm 1. The CSI of each of the N_{BS} interference channels

Algorithm 1 Channel-selection algorithm

loop
  for $i = 1 : N_{\text{BS}}$ **do**
    Estimate CSI of $\mathbf{H}_i$.
    Send $\mathbf{H}_i$ to Algorithm 2 for null space computation.
    Receive projection matrix $\mathbf{P}_{\mathbf{V}_i}$ from Algorithm 2.
  end for
  Find $i_{\min} = \arg\min_{1 \le i \le N_{\text{BS}}} \|\mathbf{x}_{\text{Radar}} - \mathbf{P}_{\mathbf{V}_i}\mathbf{x}_{\text{Radar}}\|$.
  Set $\check{\mathbf{H}} = \mathbf{H}_{i_{\min}}$ as the best interference channel.
  Set $\mathbf{P}_{\check{\mathbf{V}}} = \mathbf{P}_{\mathbf{V}_{i_{\min}}}$.
  Send $\mathbf{P}_{\check{\mathbf{V}}}$ to Algorithm 2 to get NSP radar waveform.
end loop

Algorithm 2 Null-space projection (NSP) algorithm

if $\mathbf{H}_i$ received from Algorithm 1 **then**
  Perform SVD on $\mathbf{H}_i$ (i.e., $\mathbf{H}_i = \mathbf{U}_i\boldsymbol{\Sigma}_i\mathbf{V}_i^H$).
  Find projection matrix $\mathbf{P}_{\mathbf{V}_i} = \mathbf{V}_i\boldsymbol{\Sigma}_i'^{M_T \times M_T}\mathbf{V}_i^H$.
  Send projection matrix $\mathbf{P}_{\mathbf{V}_i}$ to Algorithm 1.
end if
if $\mathbf{P}_{\check{\mathbf{V}}}$ received from Algorithm 1 **then**
  Get NSP radar signal via $\check{\mathbf{x}}_{\text{Radar}} = \mathbf{P}_{\check{\mathbf{V}}}\mathbf{x}_{\text{Radar}}$.
end if

is first estimated using a blind null-space learning algorithm [17]. Algorithm 2 gets the CSI estimates of the interference channels from Algorithm 1 and finds the null space of each $\mathbf{H}_i^{N_R^{\text{BS}} \times M_T}$. This is performed using the singular value decomposition (SVD) theorem as shown in our NSP algorithm (Algorithm 2). The SVD for the complex ith interference channel is given by

$$\mathbf{H}_i^{N_R^{\text{BS}} \times M_T} = \mathbf{U}_i\boldsymbol{\Sigma}_i^{N_R^{\text{BS}} \times M_T}\mathbf{V}_i^H$$

and $\boldsymbol{\Sigma}_i^{N_R^{\text{BS}} \times M_T}$ is given by

$$\boldsymbol{\Sigma}_i^{N_R^{\text{BS}} \times M_T} = \text{diag}(\sigma_1, \ldots, \sigma_k, \ldots, \sigma_l) \in \mathbf{R}^{N_R^{\text{BS}} \times M_T}$$

$$\text{s.t.} \quad l = \min\{N_R^{\text{BS}}, M_T\}$$

$$\sigma_i = \begin{cases} \sigma_i, & i \le k \\ 0\,, & i > k \end{cases}$$

where $\mathbf{U}_i$ is the complex unitary matrix, $\boldsymbol{\Sigma}_i$ is the matrix of singular values, $\sigma_1 > \sigma_2 > \cdots > \sigma_k > \sigma_{k+1} = \cdots = \sigma_l = 0$, and $\mathbf{V}_i^H$ is the complex unitary matrix. Once the null space of all interference channels is determined, $\boldsymbol{\Sigma}_i'^{M_T \times M_T}$ is then calculated as follows:

$$\boldsymbol{\Sigma}_i'^{M_T \times M_T} = \text{diag}(\sigma_1', \ldots, \sigma_{M_T}') \in \mathbb{R}^{M_T \times M_T}$$
$$\text{s.t.} \quad \sigma_i' = \begin{cases} 0, & i \le k \\ 1, & i > k. \end{cases}$$

Algorithm 2 uses $\boldsymbol{\Sigma}_i'^{M_T \times M_T}$ for the formation of the projection matrix $\mathbf{P}_{\mathbf{V}_i}$ that is given by

$$\mathbf{P}_{\mathbf{V}_i} = \mathbf{V}_i \boldsymbol{\Sigma}_i'^{M_T \times M_T} \mathbf{V}_i^H$$

where $\mathbf{P}_{\mathbf{V}_i}$ satisfies the following properties:

- $\mathbf{H}_i \mathbf{P}_{\mathbf{V}_i} = 0$.
- $\mathbf{P}_{\mathbf{V}_i}{}^2 = \mathbf{P}_{\mathbf{V}_i}$.

Algorithm 1 receives the projection matrices $\mathbf{P}_{\mathbf{V}_i}$ and uses them to determine the best interference channel $\breve{\mathbf{H}}$ and its corresponding $\mathbf{P}_{\breve{\mathbf{V}}}$, the one with the minimum $\|\mathbf{x}_{\text{Radar}} - \mathbf{P}_{\mathbf{V}_i}\mathbf{x}_{\text{Radar}}\|$, which according to our Algorithm 1 is given by

$$\begin{aligned} i_{\min} &= \underset{1 \le i \le N_{\text{BS}}}{\arg\min} \|\mathbf{x}_{\text{Radar}} - \mathbf{P}_{\mathbf{V}_i}\mathbf{x}_{\text{Radar}}\| \\ \breve{\mathbf{H}} &= \mathbf{H}_{i_{\min}} \\ \mathbf{P}_{\breve{\mathbf{V}}} &= \mathbf{P}_{\mathbf{V}_{i_{\min}}}. \end{aligned}$$

Algorithm 1 sends $\mathbf{P}_{\breve{\mathbf{V}}}$ to Algorithm 2 where it is used for the projection of the radar waveform. The radar waveform projected onto the null space of $\breve{\mathbf{H}}$ can be written as

$$\breve{\mathbf{x}}_{\text{Radar}} = \mathbf{P}_{\breve{\mathbf{V}}}\mathbf{x}_{\text{Radar}}. \tag{6.1}$$

6.4 RA with CA for Radar-LTE Spectrum Sharing

Each of the N_{BS} LTE-Advanced base stations has L^{UE} UEs/mobiles and two carriers. One of the carriers is the LTE-Advanced carrier that is considered to be the primary carrier and the other one is the MIMO radar carrier considered to be the secondary carrier. Each user is allocated certain bandwidth r_i based on the type of application the UE is running. Our goal is to determine the optimal bandwidth that needs to be allocated to each user by the two carriers.

Each UE has its own utility function $U_i(r_i)$ that corresponds to the application running on the UE [18–27]. We assume that the utility function assigned to the ith user is a strictly concave utility function if the user is running delay-tolerant application or a sigmoidal-like utility function if the user is running real-time application, as in [18, 27–34].

The first resource allocation optimization problem is the primary carrier (LTE-Advanced carrier) optimization. The primary carrier allocates its resources using a utility proportional fairness approach to guarantee that no user is allocated zero resources.

The LTE-Advanced carrier optimization problem can be written as:

$$\begin{aligned} \max_{\mathbf{r}_{\text{LTE}}} \quad & \prod_{i=1}^{L^{\text{UE}}} U_i(r_{i,\text{LTE}}) \\ \text{subject to} \quad & \sum_{i=1}^{L^{\text{UE}}} r_{i,\text{LTE}} \leq R_{\text{LTE}} \\ & 0 \leq r_i \leq R_{\text{LTE}}, \quad i = 1, 2, \ldots, L^{\text{UE}}. \end{aligned} \tag{6.2}$$

where $\mathbf{r}_{\text{LTE}} = \{r_{1,\text{LTE}}, r_{2,\text{LTE}}, \ldots, r_{L^{\text{UE}},\text{LTE}}\}$ and L^{UE} is the number of mobile users in the coverage area of primary carrier and R_{LTE} is the maximum achievable rate of the primary carrier. This resource allocation objective function is to maximize the total system utility when allocating resources to each user. Furthermore, it provides a proportional fairness among utilities. Users running real-time applications are allocated more resources in this approach.

The objective function in the optimization problem (6.2) is equivalent to $\max_{\mathbf{r}_{\text{LTE}}} \sum_{i=1}^{L^{\text{UE}}} \log U_i(r_{i,\text{LTE}})$, so the optimization problem (6.2) is a convex optimization problem and there exists a unique tractable global optimal solution as shown in [19]. The solution of this optimization problem is the first optimal solution that gives each of the L^{UE} users the optimal rate $r_{i,\text{LTE}}^{\text{opt}}$ only from the primary carrier and not yet the final optimal rate.

As mentioned before, once the LTE-Advanced carrier finishes allocating its resources to the L^{UE} users, the MIMO radar carrier starts to allocate its available resources to the same users using proportional fairness approach to ensure a minimum user QoS.

The optimization problem for the secondary carrier (MIMO radar) can be written as:

$$\begin{aligned} \max_{\mathbf{r}_{\text{radar}}} \quad & \prod_{i=1}^{L^{\text{UE}}} U_i\left(r_{i,\text{radar}} + r_{i,\text{LTE}}^{\text{opt}}\right) \\ \text{subject to} \quad & \sum_{i=1}^{L^{\text{UE}}} r_{i,\text{radar}} \leq R_{\text{radar}} \\ & 0 \leq r_{i,\text{radar}} \leq R_{\text{radar}}, \quad i = 1, 2, \ldots, L^{\text{UE}}. \end{aligned} \tag{6.3}$$

where $\mathbf{r}_{\text{radar}} = \{r_{1,\text{radar}}, r_{2,\text{radar}}, \ldots, r_{L_{\text{UE}},\text{radar}}\}$ and L^{UE} is the number of UEs in the coverage area, R_{radar} is the maximum achievable rate by the secondary carrier, and $r_{i,\text{LTE}}^{\text{opt}}$ is the optimal rate allocated to user i by the LTE-Advanced carrier in (6.2). Optimization problem (6.3) ensures a minimum rate of $r_{i,\text{LTE}}^{\text{opt}}$ for each user and gives priority for users running real-time applications.

The objective function in the optimization problem (6.3) is equivalent to $\max_{\mathbf{r}_{\text{radar}}} \sum_{i=1}^{L^{\text{UE}}} \log U_i(r_{i,\text{radar}} + r_{i,\text{LTE}}^{\text{opt}})$, so the optimization problem (6.3) is a convex optimization problem [35–37] and there exists a unique tractable global optimal solution [19].

The final optimal aggregated rate $r_{i,\text{agg}}$ for user i is obtained by the sum of the solution of the optimization problem (6.2) $r_{i,\text{LTE}}^{\text{opt}}$ and the solution of (6.3) $r_{i,\text{radar}}^{\text{opt}}$ and can be written as $r_{i,\text{agg}}^{\text{opt}} = r_{i,\text{radar}}^{\text{opt}} + r_{i,\text{LTE}}^{\text{opt}}$, such that $r_{i,\text{agg}}^{\text{opt}}$ is the global final optimal solution that gives each of the L^{UE} users the optimal rate from both the LTE-Advanced and the MIMO radar carriers. The solution of the optimization problem (6.3) is the global optimal solution that gives each of the M users optimal rates from both the primary and secondary carriers.

6.5 Two-Stage Carrier Aggregation Algorithm

Our two-stage algorithm is a modified version of the algorithm proposed in [38]. Algorithms 3 and 4 are the first-stage algorithms whereas Algorithms 5 and 6 are the second-stage algorithms. In the first stage, the UEs and the primary carrier collaborate to allocate an optimal rate to each UE. The first stage of the algorithm starts when each UE transmits an initial bid $w_{i,\text{LTE}}(1)$ to the LTE-Advanced eNodeB. The eNodeB checks the difference between the current received bid and the previous one, if it is less than a threshold ε it exits. Otherwise, if the difference is greater than ε, the shadow price $P_{\text{LTE}}(n) = \frac{\sum_{i=1}^{L^{\text{UE}}} w_{i,\text{LTE}}(n)}{R_{\text{LTE}}}$ is calculated by the LTE-Advanced eNodeB. The shadow price represents the total price per unit bandwidth for all users. It depends on the users bids and the eNodeB's available resources. The LTE-Advanced eNodeB sends the calculated $P_{\text{LTE}}(n)$ to each UE where it is used to calculate the rate $r_{i,\text{LTE}}(n)$ that is the solution of the optimization problem $r_{i,\text{LTE}}(n) = \arg\max_{r_{i,\text{LTE}}}(\log U_i(r_{i,\text{LTE}}) - P_{\text{LTE}}(n) r_{i,\text{LTE}})$. The calculated rate is then used to estimate a new bid $w_{i,\text{LTE}}(n)$ where $w_{i,\text{LTE}}(n) = P_{\text{LTE}}(n) r_{i,\text{LTE}}(n)$. All UEs check the fluctuation condition and send their new bids $w_{i,\text{LTE}}(n)$ to the LTE eNodeB. Once the first stage is finalized by the eNodeB, each UE calculates its allocated rate $r_{i,\text{LTE}}^{\text{opt}} = \frac{w_{i,\text{LTE}}(n)}{P_{\text{LTE}}(n)}$.

After allocating rates $r_{i,\text{LTE}}^{\text{opt}}$ from the LTE carrier, the second stage of the algorithm starts performing. Each UE transmits its initial bid $w_{i,\text{radar}}(1)$ to the MIMO radar eNodeB. The eNodeB checks the difference between the current received bid and the previous one if it is less than a threshold ε it exits. Otherwise, if the

Algorithm 3 UE first-stage algorithm

```
Send initial bid w_{i,LTE}(1) to LTE-Advanced eNodeB
loop
    Receive shadow price P_LTE(n) from LTE eNodeB
    if STOP from LTE eNodeB then
        Calculate allocated rate r_{i,LTE}^{opt} = w_{i,LTE}(n) / P_LTE(n)
    else
        Solve r_{i,LTE}(n) = arg max_{r_{i,LTE}} ( log U_i(r_{i,LTE}) - P_LTE(n) r_{i,LTE} )
        Calculate new bid w_{i,LTE}(n) = P_LTE(n) r_{i,LTE}(n)
        if |w_{i,LTE}(n) - w_{i,LTE}(n-1)| > Δw then
            w_{i,LTE}(n) = w_{i,LTE}(n-1) + sign(w_{i,LTE}(n) - w_{i,LTE}(n-1)) Δw(n)
            {Δw(n) = l_1 e^{-n/l_2}}
        end if
        Send new bid w_{i,LTE}(n) to eNodeB
    end if
end loop
```

Algorithm 4 LTE eNodeB algorithm

```
loop
    Receive bids w_{i,LTE}(n) from UEs {Let w_{i,LTE}(0) = 0 ∀i}
    if |w_{i,LTE}(n) - w_{i,LTE}(n-1)| < ε ∀i then
        STOP and allocate rates (i.e., r_{i,LTE}^{opt} to user i)
    else
        Calculate P_LTE(n) = Σ_{i=1}^{L_UE} w_{i,LTE}(n) / R_LTE
        Send new shadow price P_LTE(n) to all UEs
    end if
end loop
```

difference is greater than ε, the MIMO radar eNodeB calculates the shadow price $P_{\text{radar}}(n) = \frac{\sum_{i=1}^{L_{\text{UE}}} w_{i,\text{radar}}(n)}{R_{\text{radar}}}$. The radar eNodeB sends the calculated $P_{\text{radar}}(n)$ to the UEs. Each UE calculates the rate $r_{i,\text{radar}}(n)$ which is the solution of the optimization problem $r_{i,\text{radar}}(n) = \arg\max_{r_{i,\text{radar}}}(\log U_i(r_{i,\text{radar}} + r_{i,\text{LTE}}^{\text{opt}}) - P_{\text{radar}}(n) r_{i,\text{radar}})$. A new bid $w_{i,\text{radar}}(n)$ is calculated using $r_{i,\text{radar}}(n)$ where $w_{i,\text{radar}}(n) = P_{\text{radar}}(n) r_{i,\text{radar}}(n)$. All UEs check the fluctuation condition and send their new bids $w_{i,\text{radar}}(n)$ to the radar eNodeB. The second stage of the algorithm is finalized by the radar eNodeB. Each UE then calculates its allocated rate $r_{i,\text{radar}}^{\text{opt}} = \frac{w_{i,\text{radar}}(n)}{P_{\text{radar}}(n)}$ by the radar eNodeB. The final global optimal rate $r_{i,\text{agg}}^{\text{opt}} = r_{i,\text{radar}}^{\text{opt}} + r_{i,\text{LTE}}^{\text{opt}}$ is then allocated to each UE. Figure 6.2 shows a flow diagram of the LTE-Advanced two-stage RA with carrier aggregation algorithm.

Algorithm 5 UE second-stage algorithm

Send initial bid $w_{i,\text{radar}}(1)$ to the radar eNodeB
loop
  Receive shadow price $P_{\text{radar}}(n)$ from the radar eNodeB
  if STOP from the radar eNodeB **then**
    Calculate allocated rate $r_{i,\text{agg}}^{\text{opt}} = \frac{w_{i,\text{radar}}(n)}{P_{\text{radar}}(n)} + r_{i,\text{LTE}}^{\text{opt}}$
  else
    Solve $r_{i,\text{radar}}(n) = \arg\max_{r_{i,\text{radar}}} \Big(\log U_i(r_{i,\text{radar}} + r_{i,\text{LTE}}^{\text{opt}}) - P_{\text{radar}}(n) r_{i,\text{radar}} \Big)$
    Calculate new bid $w_{i,\text{radar}}(n) = P_{\text{radar}}(n) r_{i,\text{radar}}(n)$
    if $|w_{i,\text{radar}}(n) - w_{i,\text{radar}}(n-1)| > \Delta w$ **then**
      $w_{i,\text{radar}}(n) = w_{i,\text{radar}}(n-1) + \text{sign}(w_{i,\text{radar}}(n) - w_{i,\text{radar}}(n-1)) \Delta w(n)$
      $\{\Delta w(n) = l_1 e^{-\frac{n}{l_2}}\}$
    end if
    Send new bid $w_{i,\text{radar}}(n)$ to eNodeB
  end if
end loop

Algorithm 6 MIMO radar eNodeB algorithm

loop
  Receive bids $w_{i,\text{radar}}(n)$ from UEs $\{\text{Let } w_{i,\text{radar}}(0) = 0 \ \forall i\}$
  if $|w_{i,\text{radar}}(n) - w_{i,\text{radar}}(n-1)| < \varepsilon \ \forall i$ **then**
    STOP and allocate rates (i.e., $r_{i,\text{radar}}^{\text{opt}}$ to user i)
  else
    Calculate $P_{\text{radar}}(n) = \frac{\sum_{i=1}^{L^{\text{UE}}} w_{i,\text{radar}}(n)}{R_{\text{radar}}}$
    Send new shadow price $P_{\text{radar}}(n)$ to all UEs
  end if
end loop

6.6 Numerical Results

In our spectrum sharing model, the LTE-Advanced system has N_{BS} BSs, only the ith BS is under zero interference from the MIMO radar due to the spectrum sharing approach employed by the proposed spectrum sharing model. We consider this BS which has two eNodeBs, one is configured at the LTE-Advanced carrier and the second is configured to use radar carrier when there is no interference from radar. In this BS we consider four UEs in its coverage area subscribing for a mobile service. The first and second UEs are running real-time applications presented by sigmoidal-like utility functions whereas the third and fourth UEs are running delay-tolerant applications presented by logarithmic utility functions. The four UEs are to be allocated resources from the LTE-Advanced and the MIMO radar carriers.

The proposed RA with CA algorithm is applied in C++ to the sigmoidal-like and logarithmic utility functions. Simulation results showed convergence to the optimal global point in the two stages of the algorithm. Each of the four UEs is allocated a final optimal rate by the two carriers. We use a normalized sigmoidal-like utility function that is expressed by Eq. (2.1) to represent the first user real-time application

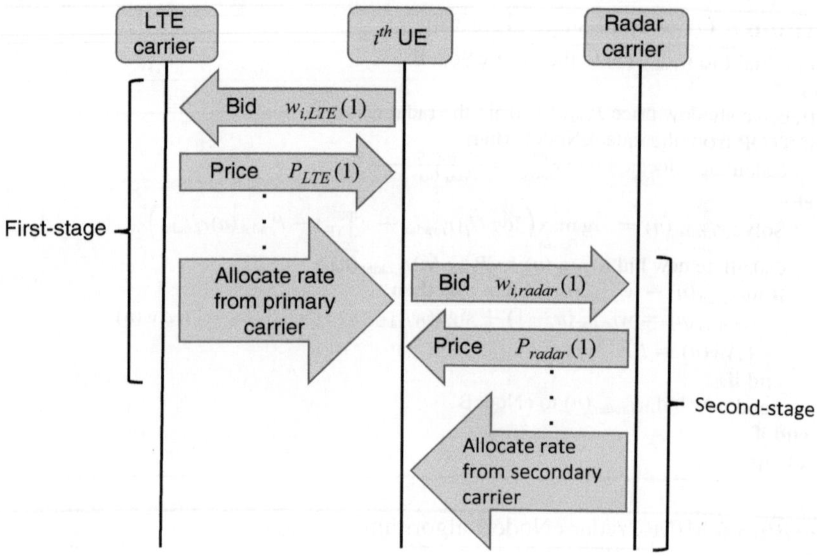

Fig. 6.2 Flow diagram for the two-stage RA with carrier aggregation algorithm for the radar-LTE spectrum sharing

with $a = 3$, $b = 20$ which is an approximation to a step function at rate $r = 20$. Additionally, we use another sigmoidal-like utility function to represent the second user real-time application with $a = 1$, $b = 30$. Furthermore, we use logarithmic functions to represent the third and fourth UEs delay-tolerant applications with $k = 3$ and $k = 0.5$, respectively. Additionally, We use $r_{\max} = 100$ for all logarithmic functions, $l_1 = 5$ and $l_2 = 10$ in the fluctuation decay function of the algorithm and $\varepsilon = 10^{-7}$. The utility functions corresponding to the four UEs applications are shown in Fig. 6.3.

6.6.1 *Gain in LTE-Advanced Cellular Network Resources Due to Spectrum Sharing with MIMO Radar*

In the following simulations, we show the total available resources R in N_{BS} LTE-Advanced cells due to spectrum sharing between the LTE-Advanced cellular network and the MIMO radar that can be expressed as

$$R = (N_{\text{BS}} - 1)\left(R_{\text{LTE}} + \left(\frac{D-1}{D}\right)R_{\text{radar}}\right) + 1(R_{\text{LTE}} + R_{\text{radar}}), \tag{6.4}$$

where R_{LTE} is one LTE eNodeB available resources (i.e., in one LTE-Advanced cellular cell), R_{radar} is the MIMO radar carrier available resources, D is the number

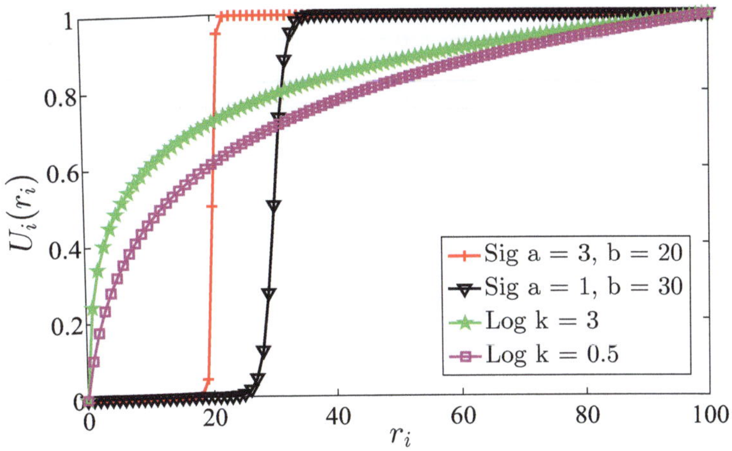

Fig. 6.3 The users utility functions $U_i(r_i)$

of segments of the radar available spectrum such that one segment is to be used by the radar and the radar can share its available $D-1$ spectrum segments with the LTE cellular cell, $\frac{D-1}{D}$ is the proportion of R_{radar} allocated to each of the $N_{\text{BS}}-1$ cellular cells, and R_{radar} is allocated to one LTE-Advanced cellular cell (i.e., with no interference to the MIMO radar signals) users.

In Fig. 6.4, we set $R_{\text{LTE}} = 70$, $R_{\text{radar}} = 80$, and $D = 3$. We show the total available resources R in N_{BS} LTE-Advanced cellular cells with different numbers of N_{BS} that varies between 5 and 20. The figure shows that R increases linearly with an increasing number of LTE-Advanced cells.

In Fig. 6.5, we set $R_{\text{LTE}} = 70$, $R_{\text{radar}} = 80$, and $N_{\text{BS}} = 10$. We show the total available resources R in 10 LTE-Advanced cellular cells with different values of D that varies between 2 and 10. The figure shows that R increases for higher $\frac{D-1}{D}$.

6.6.2 *Rate Allocation for* $10 \leq R_{\text{LTE}} \leq 70$ *in the First Stage of the RA Algorithm*

We apply Algorithms 3 and 4 of the first stage in C++ to the sigmoidal-like and logarithmic utility functions. The LTE-Advanced eNodeB available resources R_{LTE} takes values between 10 and 70 with step of 10. In Fig. 6.6, we show the four users optimal rates $r_{i,\text{LTE}}^{\text{opt}}$ allocated by the LTE-Advanced eNodeB with different eNodeB resources R_{LTE}. This represents the solution of optimization problem (6.2). As mentioned before the sigmoidal-like utility functions are given priority over the logarithmic utility functions for rate allocation and this explains the results we got in Fig. 6.6 where the algorithm gives priority to real-time applications when allocating the LTE-Advanced eNodeB resources as it uses proportional fairness

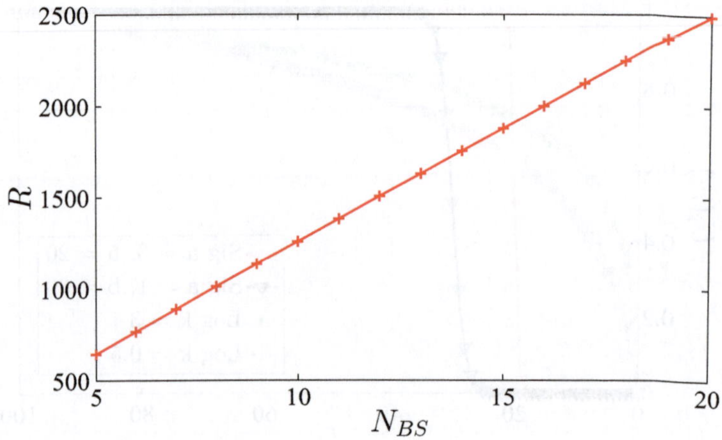

Fig. 6.4 The total LTE-Advanced network available resources R in N_{BS} cellular cells due to spectrum sharing between LTE-Advanced and MIMO radar carriers when $R_{\text{LTE}} = 70$ and $R_{\text{radar}} = 80$ with different numbers of LTE-Advanced cellular cells when $\frac{2}{3}R_{\text{radar}}$ is allocated to $N_{\text{BS}} - 1$ and R_{radar} is allocated to one LTE-Advanced cellular cell (i.e., with no interference to the MIMO radar signals)

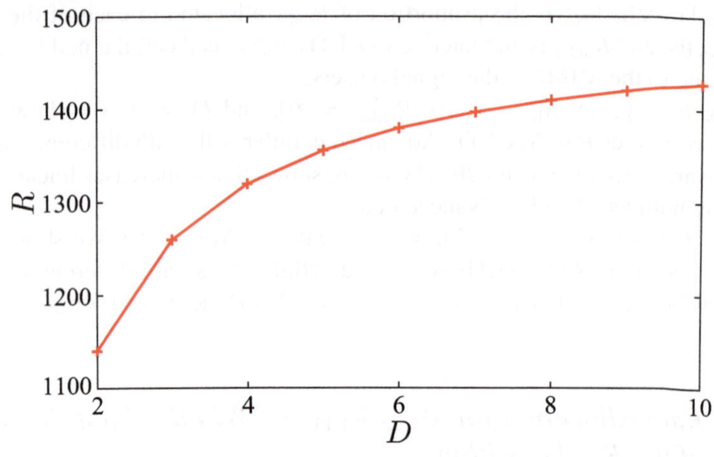

Fig. 6.5 The total LTE-Advanced network available resources R in 10 cellular cells due to spectrum sharing between LTE-Advanced and MIMO radar carriers with $R_{\text{LTE}} = 70$ and $R_{\text{radar}} = 80$ with different values of D where $\frac{D-1}{D}$ is the proportion of R_{radar} allocated to nine cellular cells users and R_{radar} is allocated to one LTE-Advanced cellular cell (i.e., with no interference to the MIMO radar signals)

approach. Users with real-time applications bid higher than the other users until each one of them reaches its inflection point then the algorithm starts dividing the remaining resources among users running delay-tolerant applications based on their utility functions parameters.

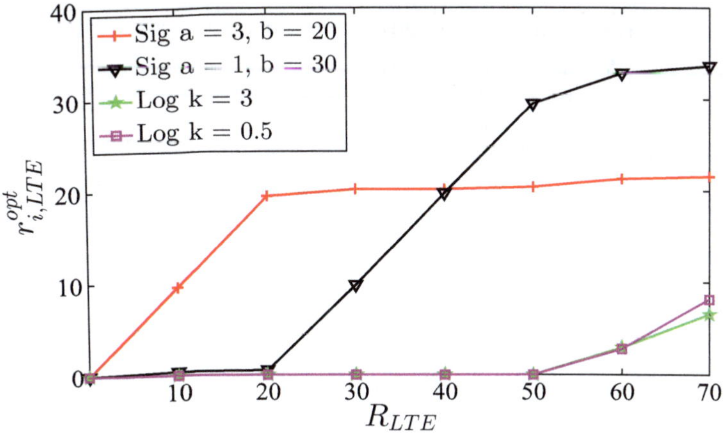

Fig. 6.6 The users optimal rates $r^{\text{opt}}_{i,\text{LTE}}$ for different values of R_{LTE} for Algorithms 3 and 4

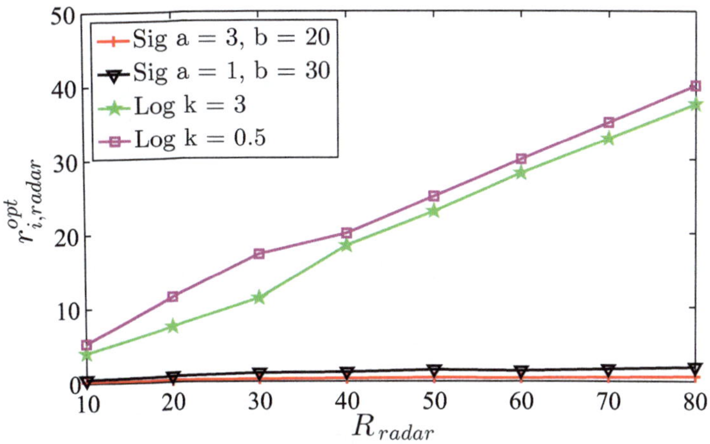

Fig. 6.7 The users optimal rates $r^{\text{opt}}_{i,\text{radar}}$ for different values of R_{radar} for Algorithms 5 and 6

6.6.3 *Rate Allocation for* $10 \leq R_{\text{radar}} \leq 80$ *in the Second Stage of the RA Algorithm*

We apply Algorithm 5 and 6 of the second stage in C++ to the sigmoidal-like and logarithmic utility functions. The radar carrier available resources R_{radar} takes values between 10 and 80 with step of 10. In Fig. 6.7, we show the four users optimal rates $r^{\text{opt}}_{i,\text{radar}}$ allocated by the radar eNodeB with different available resources R_{radar}. This represents the solution of optimization problem (6.3). Each user running real-time application is allocated at least its utility inflection rate $r_i = b_i$ by the LTE-Advanced

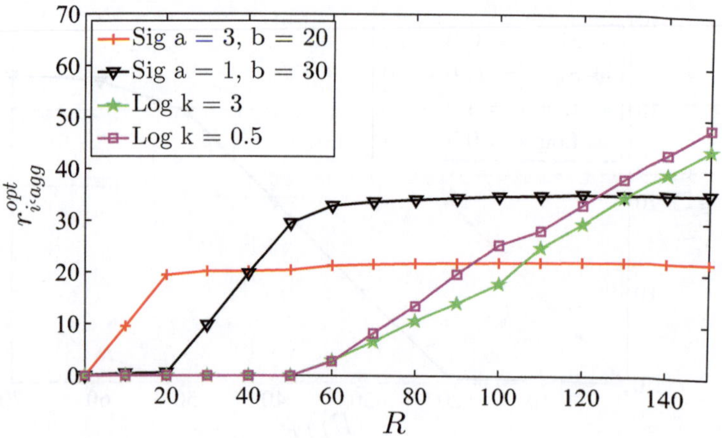

Fig. 6.8 The users final optimal rates $r_{i,\text{agg}}^{\text{opt}}$ for different values of R where $10 \leq R \leq 70$ is the LTE-Advanced carrier available resources and $70 < R \leq 150$, R is the total available resources of $R_{\text{LTE}} = 70$ and $10 \leq R_{\text{radar}} \leq 80$

carrier in the first stage of the algorithm, this explains the result we got in Fig. 6.7 where most of the radar carrier resources are allocated to users running delay-tolerant applications.

6.6.4 *RA with Carrier Aggregation for* $10 \leq R \leq 150$

In the following simulations, the total rate of the LTE-Advanced carrier takes values between 10 and 70 and the MIMO radar carrier has available resources that takes values between 10 and 80. The two carriers resources are to be allocated to the four users subscribing for a mobile service in the LTE-Advanced cellular cell using RA with carrier aggregation.

In Fig. 6.8, we show the optimal rate allocated to each user by the first stage of the algorithm when $10 \leq R \leq 70$ is the LTE-Advanced carrier available resources. The final optimal rates allocated to each user by the second stage of the algorithm are also shown in Fig. 6.8 for $70 < R \leq 150$ where R is the total available resources of $R_{\text{LTE}} = 70$ and $10 \leq R_{\text{radar}} \leq 80$. The LTE carrier allocates the majority of its resources to the UEs running real-time applications until they reach the inflection rate $r_i = b_i$. When the LTE-Advanced resources R_{LTE} exceed the total inflection rates of the users real-time applications, the LTE-Advanced carrier starts allocating resources to the delay-tolerant applications. The aggregated final optimal rate allocated to each user by the LTE-Advanced and the radar carriers is the total optimal rate allocated to each user by the LTE-Advanced carrier when $R_{\text{LTE}} = 70$ and the optimal rate allocated to the same user by the radar carrier when $10 \leq R_{\text{radar}} \leq 80$. Since users running real-time applications are allocated at least

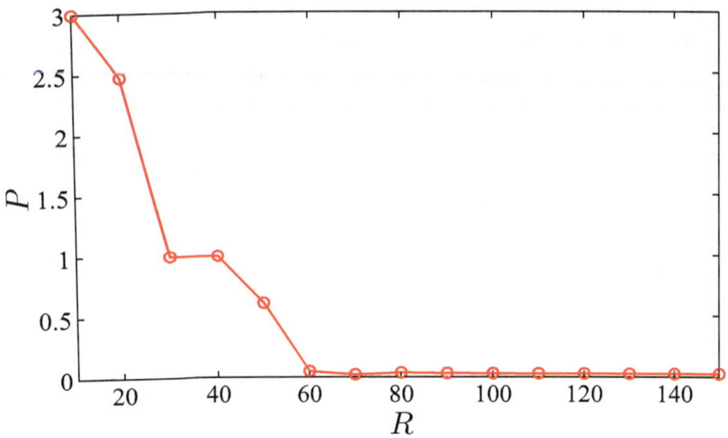

Fig. 6.9 The shadow price P for different values of R and fixed number of users (same four users), R is the LTE-Advanced carrier available resources for $10 \leq R \leq 70$ whereas when $70 < R \leq 150$, R is the total available resources of $R_{\text{LTE}} = 70$ and $10 \leq R_{\text{radar}} \leq 80$

their utilities inflection rates $r_i = b_i$ by the LTE-Advanced carrier, the radar carrier allocates most of its resources to users running delay-tolerant applications.

6.6.5 *Price Sensitivity to Change in R*

In the following simulations, the total available resources take different values between 10 and 150 with step of 10. In Fig. 6.9, we show the shadow price P, that represents the total price per unit bandwidth for all users, with the total available resources R of the LTE-Advanced and radar carriers. R is the LTE-Advanced carrier available resources for $10 \leq R \leq 70$ whereas when $70 < R \leq 150$, R is the total available resources of $R_{\text{LTE}} = 70$ and $10 \leq R_{\text{radar}} \leq 80$. As expected the price is higher for smaller R when the number of users is fixed (the same four users).

6.7 Summary and Conclusion

In this chapter, we presented a spectrum sharing scenario between a MIMO radar and LTE cellular system with multiple BSs. We proposed a channel-selection algorithm and NSP algorithm to select the best interference channel and project the radar signal onto it. Our proposed algorithms guarantee a minimum degradation in the radar's performance by selecting the best interference channel for the NSP of the radar signal. In addition, we presented an optimal resource allocation with carrier aggregation approach to allocate LTE-Advanced and MIMO radar carriers' resources optimally among LTE-Advanced users in a cellular cell. We considered two utility functions based on the application type running on the UE, sigmoidal-like utility functions represent real-time applications and logarithmic utility functions represent delay-tolerant applications. As a result of our analysis, we presented an iterative distributed RA with carrier aggregation algorithm for the UEs and both LTE-Advanced and radar carriers. The algorithm provides a utility proportional fair resource allocation which guarantees a minimum QoS to each user while giving priority to users running real-time applications. We showed through simulations that our algorithm converges to the optimal rates in its two stages.

References

1. G. Yuan, X. Zhang, W. Wang, Y. Yang, Carrier aggregation for LTE-advanced mobile communication systems. IEEE Commun. Mag. **48**, 88–93 (2010)
2. Z. Shen, A. Papasakellariou, J. Montojo, D. Gerstenberger, F. Xu, Overview of 3GPP LTE-advanced carrier aggregation for 4G wireless communications. IEEE Commun. Mag. **50**, 122–130 (2012)
3. *Frequency Spectrum Wall Chart* (Commerce Dept., National Telecommunications and Information Administration, Office of Spectrum Management, 2016)
4. National Telecommunications and Information Administration (NTIA), An assessment of the near-term viability of accommodating wireless broadband systems in the 1675–1710 MHz, 1755–1780 MHz, 3500–3650 MHz, 4200–4220 MHz, and 4380–4400 MHz bands (Fast Track Report), Online, October 2010
5. NTIA, Department of Commerce, An assessment of the near-term viability of accommodating wireless broadband systems in the 1675–1710MHz, 1755–1780MHz, 3500–3650MHz, 4200–4220MHz and 4380–4400MHz bands. Technical Report (2010)
6. National Telecommunications and Information Administration (NTIA), Analysis and resolution of RF interference to radars operating in the band 2700–2900 MHz from broadband communication transmitters, Online, October 2012
7. M. Cotton, R. Dalke, Spectrum occupancy measurements of the 3550–3650 MHz maritime radar band near San Diego, California. Report (2014)
8. Executive Office of the President, President's Council of Advisors on Science and Technology (PCAST), *Realizing the Full Potential of Government-Held Spectrum to Spur Economic Growth* (PCAST, Washington, DC, 2012)
9. M. Richards, J. Scheer, W. Holm, *Principles of Modern Radar* (Scitech, New York, 2010)
10. Federal Communications Commission (FCC), In the matter of revision of parts 2 and 15 of the commission's rules to permit unlicensed national information infrastructure (U-NII) devices in the 5 GHz band. MO and O, ET Docket No. 03–122, June 2006
11. Federal Communications Commission, Proposal to create a Citizen's broadband service in the 3550–3650 MHz band. Report (2012)
12. Federal Communications Commission (FCC), Connecting America: The national broadband plan (Online, 2010)
13. A. Abdelhadi, T.C. Clancy, Network MIMO with partial cooperation between radar and cellular systems, in *2016 International Conference on Computing, Networking and Communications (ICNC)* (2016), pp. 1–5
14. J. Li, P. Stoica, MIMO radar with colocated antennas. IEEE Signal Process. Mag. **24**(5), 106–114 (2007)

15. H. Shajaiah, A. Abdelhadi, C. Clancy, Spectrum sharing approach between radar and communication systems and its impact on radar's detectable target parameters, in *2015 IEEE 81st Vehicular Technology Conference (VTC Spring)* (2015), pp. 1–6
16. H. Shajaiah, A. Khawar, A. Abdel-Hadi, T. Clancy, Resource allocation with carrier aggregation in LTE Advanced cellular system sharing spectrum with S-band radar, *2014 IEEE International Symposium on Dynamic Spectrum Access Networks (DYSPAN)* (2014), pp. 34–37
17. Y. Noam, A. Goldsmith, Blind null-space learning for mimo underlay cognitive radio with primary user interference adaptation. IEEE Trans. Wirel. Commun. **12**(4), 1722–1734 (2013)
18. G. Tychogiorgos, A. Gkelias, K.K. Leung, Utility-proportional fairness in wireless networks, in *International Symposium on Personal, Indoor, and Mobile Radio Communications* (IEEE, New York, 2012), pp. 839–844
19. A. Abdelhadi, C. Clancy, A utility proportional fairness approach for resource allocation in 4G-LTE, in *IEEE International Conference on Computing, Networking, and Communications (ICNC), CNC Workshop* (2014)
20. A. Abdelhadi, C. Clancy, A robust optimal rate allocation algorithm and pricing policy for hybrid traffic in 4G-LTE, in *IEEE International Symposium on Personal, Indoor, and Mobile Radio Communications (PIMRC)* (2013)
21. Y. Wang, A. Abdelhadi, A QoS-based power allocation for cellular users with different modulations, in *2016 International Conference on Computing, Networking and Communications (ICNC)* (2016), pp. 1–5
22. A. Abdelhadi, H. Shajaiah, Optimal resource allocation for smart phones with multiple applications with MATLAB instructions. Technical Report (2016)
23. A. Abdelhadi, H. Shajaiah, Application-aware resource allocation with carrier aggregation using MATLAB. Technical Report (2016)
24. Z. Kbah, A. Abdelhadi, Resource allocation in cellular systems for applications with random parameters, in *2016 International Conference on Computing, Networking and Communications (ICNC)* (2016), pp. 1–5
25. A. Abdelhadi, H. Shajaiah, Optimal resource allocation for cellular networks with MATLAB instructions. CoRR, abs/1612.07862 (2016)
26. F. Wilson, I. Wakeman, W. Smith, Quality of service parameters for commercial application of video telephony, in *Proceedings of the Human Factors in Telecommunications Symposium* (Darmstadt, 1993), pp. 139–148
27. S. Shenker, Fundamental design issues for the future internet. IEEE J. Sel. Areas Commun. **13**, 1176–1188 (1995)
28. Y. Wang, A. Abdelhadi, T. C. Clancy, Optimal power allocation for LTE users with different modulations, in *2016 Annual IEEE Systems Conference (SysCon)* (2016), pp. 1–5
29. M. Ghorbanzadeh, A. Abdelhadi, C. Clancy, Distributed resource allocation, in *Cellular Communications Systems in Congested Environments* (Springer, Berlin, 2017), pp. 61–91
30. J.-W. Lee, R.R. Mazumdar, N.B. Shroff, Downlink power allocation for multi-class wireless systems. IEEE/ACM Trans. Netw. **13**, 854–867 (2005)
31. A. Kumar, A. Abdelhadi, T.C. Clancy, A delay efficient multiclass packet scheduler for heterogeneous M2M uplink, in *IEEE MILCOM* (2016)
32. A. Kumar, A. Abdelhadi, T.C. Clancy, An online delay efficient packet scheduler for M2M traffic in industrial automation, in *IEEE Systems Conference* (2016)
33. A. Kumar, A. Abdelhadi, T.C. Clancy, A delay optimal MAC and packet scheduler for heterogeneous M2M uplink. CoRR, abs/1606.06692 (2016)
34. A. Kumar, A. Abdelhadi, T.C. Clancy, A delay-optimal packet scheduler for M2M uplink, in *IEEE MILCOM* (2016)
35. R. Madan, S. Boyd, S. Lall, Fast algorithms for resource allocation in wireless cellular networks. IEEE/ACM Trans. Netw. **18**, 973–984 (2010)
36. J.B. Taylor, *Principles of Microeconomics*. Microeconomics Series (Houghton Mifflin, Boston, 1998)

37. S. Boyd, L. Vandenberghe, *Introduction to Convex Optimization with Engineering Applications*. Convex Optimization, Course Reader for EE364 (1999)
38. H. Shajaiah, A. Abdelhadi, C. Clancy, Utility proportional fairness resource allocation with carrier aggregation in 4G-LTE, in *IEEE Military Communications Conference (MILCOM)* (2013)

Chapter 7
Utility Proportional Fairness Resource Block Scheduling with Carrier Aggregation

Resource allocation for multi-carrier systems in single cell have been given attention in recent years [1–7]. Utility proportional fairness resource allocation for a single carrier in cellular networks have been extensively studied in [8–10]. The authors in [1, 11–13] have presented multi-stage resource allocation with carrier aggregation algorithms for multi-carrier cellular systems. However, none of their resource allocation approaches have considered the problem of resource block scheduling for multiple component carriers. This problem is an another essential step for addressing the recommendations of President Council of Advisers on Science and Technology report [14], Federal Communications Commission (FCC) [15–18], and the National Telecommunications and Information Administration (NTIA) [19–21] to aggregate federal spectrum with commercial.

In this chapter, we focus on solving the problem of utility PF resource block scheduling with CA for multi-carrier cellular networks. The resource scheduling approach presented in [22, 23] does not consider the case of multi-carrier resources available at the eNodeB. It only solves the problem of RB scheduling in the case of single carrier. We introduce an approach for resource block scheduling with carrier aggregation in LTE-Advanced cellular networks. In our model, users are running elastic or inelastic traffic. We use logarithmic and sigmoidal-like utility functions to represent the applications running on the user equipment. Our objective is to assign resource blocks from multiple carriers based on a proportional fairness scheduling policy. In our approach, users are partitioned into different groups based on the carriers coverage area. In each group of users, the UEs are assigned RBs from all in band carriers. We use a utility proportional fairness approach in the utility percentage of the application running on the UE. Each user is guaranteed a minimum

---

The content in this chapter is reproduced with permission after modifications (License number 4078221405942). For the original article please refer to [24].

H. Shajaiah et al., *Resource Allocation with Carrier Aggregation in Cellular Networks*, DOI 10.1007/978-3-319-60540-1_7

quality of service with a priority criterion that is based on the type of application running on the UE. The contributions in this chapter are summarized as:

- We propose a framework for the problem of utility proportional fairness RB scheduling with CA for multi-carrier cellular networks.
- We introduce a user grouping method that creates a user group for each component carrier such that each carrier assigns its resources only to users in its user group. Each user subscribing for a mobile service is assigned on multiple component carriers' RBs based on the proposed user grouping method and a utility proportional fairness policy.
- We prove that the proposed resource scheduling policy, that is based on CA, exists and that the optimal solution is tractable.
- We present a centralized utility proportional fairness resource scheduling with carrier aggregation (UPFRS-CA) algorithm to allocate multiple carriers resources optimally among users.
- We present simulation results for the proposed resource scheduling with CA approach and compare its performance in the case of using the proposed resource scheduling policy and the case of using the scheduling policy presented in [23].

7.1 System Model and Problem Setup

The transmission resources in an LTE downlink have dimensions in frequency, time, and space [25]. The frequency is represented by subcarriers. The time is divided into frames and each frame is further divided into subframes. The space is provided by the transmit and receive antennas. One RB consists of 12 continuous subcarriers. In reuse-1 radio systems, that is considered in this chapter, a RB can be allocated to only one user.

We consider a single cell LTE-Advanced mobile system with one eNodeB and M users. Let the number of CCs that the system can aggregate be K. The set of CCs is given by $\mathcal{K} = \{f_1, f_2, \ldots, f_K\}$ with CCs in order from the highest frequency to the lowest frequency (i.e., $f_1 > f_2 > \cdots > f_K$). We consider an equal power allocation (EPA) scheme that each frequency component has the same transmitting power. Furthermore, a non-adjacent inter-band aggregation scenario is considered. Because the channel fading for high frequency is larger than that for low frequency, higher frequency carriers have smaller coverage areas than lower frequency carriers. Users located under the coverage area of multiple carriers are scheduled resources from all in band carriers. The eNodeB assigns RBs from multiple carriers to each UE. The total allocated rate achieved by assigning RBs to the ith UE is given by r_i. Each UE has its own utility function $U_i(r_i)$ that corresponds to the type of application running on the ith UE. Our goal is to determine which RBs from each CC should be allocated to each UE by the eNodeB in order to maximize the total system utility while ensuring PF between utilities.

We define $\mathscr{Z}_k$, where $1 \leq k \leq K$, to be the set of RBs available by f_k carrier where $z_{k,j}$ denotes a single RB in $\mathscr{Z}_k = \{z_{k,1}, z_{k,2}, \ldots\}$, $z_{k,j} \in \mathscr{Z}_k$ is the jth RB in CC f_k and $|\mathscr{Z}_k|$ denotes the number of RBs available by f_k carrier. The signal to noise ratio (SNR) of user i on RB $z_{k,j}$ is given by

$$\gamma_{i,z_{k,j}} = P_{z_{k,j}}|G_{i,z_{k,j}}|^2/N_{i,z_{k,j}}, \tag{7.1}$$

where $G_{i,z_{k,j}}$ is the complex channel gain between the eNodeB and the ith UE on RB $z_{k,j}$, $N_{i,z_{k,j}}$ is the noise power experienced by the ith UE on RB $z_{k,j}$, and $P_{z_{k,j}}$ is the transmission power that the eNodeB assigns to RB $z_{k,j}$. Under the EPA, $P_{z_{k,j}} = P_k/|\mathscr{Z}_k|$ where P_k is the transmitting power of CC f_k. Then the achievable data rate of the ith user on RB $z_{k,j}$ is given by

$$H_{i,z_{k,j}} = W \log(1 + \beta_{z_{k,j}} \gamma_{i,z_k}), \tag{7.2}$$

where W is the bandwidth of a RB and $\beta_{z_{k,j}}$ is the SNR gap, between the practical implementations and the information theoretical results, that is needed to reach certain capacity.

In each frame, the eNodeB schedules each of the frame's RBs to one UE. Let $\phi_{i,z_{k,j}}$ be the proportion of frames that the ith UE is scheduled by the eNodeB on RB $z_{k,j}$. The ith UE rate on all RBs scheduled by carrier f_k is given by

$$r_{i,f_k} = \sum_{z_{k,j} \in \mathscr{Z}_k} \phi_{i,z_{k,j}} H_{i,z_{k,j}}. \tag{7.3}$$

The overall rate of the ith UE, that is the sum of the rates achieved by all carriers RBs assignments, is given by $r_i = \sum_{f_k \in \mathscr{K}} r_{i,f_k}$.

We express the user satisfaction with its application rates using utility functions [8, 9, 26–33]. We represent the ith user application utility function $U_i(r_i)$ by sigmoidal-like function or logarithmic function where r_i is the rate of the ith user application. Logarithmic utility functions expressed by Eq. (2.2) and sigmoidal-like utility functions expressed by Eq. (2.1) are used to represent delay-tolerant and real-time applications, respectively, as in [26, 33–40]

A user grouping method is introduced in Sect. 7.2 to partition users into groups depending on their location in the cell. The eNodeB performs RBs assignments from each CC to the user group located in the coverage area of that carrier.

7.2 User Grouping Method

In this section we introduce a user grouping method to create one user group $\mathscr{M}_{f_k}$ for each CC f_k where $\mathscr{M}_{f_k}$ is a set of users located under the coverage area of carrier f_k. Users in $\mathscr{M}_{f_k}$ are assigned RBs on CC f_k by the eNodeB. Users located under the coverage area of multiple carriers (i.e., common users in multiple user groups) are assigned RBs on these carriers and their final rates are aggregated under a non-adjacent inter-band aggregation scenario.

The ith user is part of user group $\mathcal{M}_{f_k}$ if it satisfies certain path loss constraints on CC f_k. Assume that the maximum path loss in a carrier cannot exceed a threshold Lth. In order for the eNodeB to identify a user group for each CC, it first computes the ith user path loss on each CC and creates a set α_i that includes all in range carriers such that the ith user is assigned RBs only from carriers in α_i.

Higher frequency carriers have smaller coverage radius R_k than lower frequency carriers (i.e., $R_1 < R_2 < \cdots < R_K$). Therefore, user group $\mathcal{M}_{f_1} \subseteq \mathcal{M}_{f_2} \subseteq \cdots \subseteq \mathcal{M}_{f_K}$.

7.3 RB Scheduling with CA Problem

In this section, we present our RB scheduling with CA approach. Our objective is to assign RBs to each user (i.e., the ith user) on all of its in range carriers (i.e., CCs in α_i) based on a utility PF policy. We use utility functions of users' applications rates to represent the type of application running on the UE. Given that different applications may have different QoS requirements, every user subscribing for a mobile service is guaranteed to achieve minimum QoS for each of its applications with a priority criterion. Users running real-time applications are given priority when assigning RBs due to the sigmoidal-like utility functions nature used to represent their applications. In addition, our utility PF approach guarantees that no user is assigned zero RBs.

The eNodeB performs the RBs assignment for each of the CC's RBs in $\mathcal{Z}_k$. It assigns the RBs of each CC f_k one at a time and one after another in ascending order of their coverage radius R_k. It starts with CC f_1 as it has the smallest coverage radius R_1. After assigning all users in $\mathcal{M}_{f_1}$ on f_1 RBs, the eNodeB then assigns users in $\mathcal{M}_{f_2}$ on f_2 RBs. In addition, since $\mathcal{M}_{f_1}$ users are also in $\mathcal{M}_{f_2}$ (i.e., $\mathcal{M}_{f_1} \subseteq \mathcal{M}_{f_2}$), the eNodeB assigns $\mathcal{M}_{f_1}$ users on f_2 RBs and the rates are aggregated based on a non-adjacent inter-band aggregation scenario. The eNodeB continues the RB assignment process by assigning $\mathcal{M}_{f_k}$ users on CC f_k RBs. Finally, the RB assignment process is finalized by assigning carrier f_K RBs to all users in the cellular cell as they are all located within its coverage radius. We consider a utility PF objective function, based on CA, that the eNodeB seeks to maximize for each time it assigns user on a RB. The utility PF resource scheduling with CA optimization problem for the eNodeB assignments of $\mathcal{M}_{f_k}$ users on $\mathcal{Z}_k$ RBs is given by

$$
\begin{aligned}
\max_{\phi_{i,z_k}} \quad & \prod_{i=1}^{M_k} U_i\Big(c_{i,f_k} + \sum_{z_{k,j}\in\mathcal{Z}_k} (\phi_{i,z_{k,j}} H_{i,z_{k,j}})\Big) \\
\text{subject to} \quad & \sum_{i=1}^{M_k} \phi_{i,z_{k,j}} = 1, \quad c_{i,f_1} = 0, \\
& \phi_{i,z_{k,j}} \geq 0, \quad i = 1, 2, \ldots, M_k \\
& c_{i,f_k} = \sum_{l=1}^{k-1} r_{i,f_l}, \quad k > 1.
\end{aligned}
\tag{7.4}
$$

where $M_k = |\mathcal{M}_{f_k}|$ is the number of UEs in the coverage area of carrier f_k, $c_{i,f_1} = 0$ and c_{i,f_k} for $k > 1$ is equivalent to $\sum_{l=1}^{k-1} r_{i,f_l}$ that is the ith UE total rate on all RBs scheduled by carriers $\{f_1, \ldots, f_{k-1}\}$. The eNodeB seeks to maximize the objective function of this resource scheduling optimization problem that is achieved by maximizing the product of all UEs' utilities when assigning the UEs on the carriers' RBs. The goal of this resource scheduling objective function is to allocate the resources to the UE that maximizes the total cellular network objective (i.e., the product of the utilities of all UEs) while ensuring PF between individual utilities. This objective function ensures non-zero RA for all users. Therefore, the resource scheduling optimization problem guarantees minimum QoS for all users. In addition, this approach allocates more resources to real-time applications providing improvement to the QoS of LTE system.

Later in this section we prove that there exists a tractable global optimal solution to optimization problem (7.4). However, the user's final rate, achieved by assigning each user on its in range carriers' RBs, is determined using a multi-stage approach where optimization problem (7.4) is required for each CC f_k. In addition, optimization problem (7.4) needs to be applied in a multi-stage scenario starting from the carrier with the smallest coverage area (i.e., f_1) and ending with the carrier that has the largest coverage area (i.e., f_K). The rate achieved for each user after assigning CC f_k RBs is needed for the next stage optimization problem (7.4) of carrier f_{k+1}. The objective function in optimization problem (7.4) is equivalent to $\arg\max_{\phi_{i,z_k}} \sum_{i=1}^{M_k} \log(U_i(c_{i,f_k} + \sum_{z_{k,j} \in \mathcal{Z}_k} (\phi_{i,z_{k,j}} H_{i,z_{k,j}})))$. The utility functions $\log(U_i(c_{i,f_k} + \sum_{z \varepsilon Z} \phi_{i,b(i),z} H_{i,b(i),z}))$ that are equivalent to $\log(U_i(c_{i,f_k} + r_{i,f_k}))$ are strictly concave functions as proved in [8]. As a result, optimization problem (7.4) is a convex optimization problem [41–43] and there exists a unique tractable global optimal solution [1, 8].

In order to consider the case when the entire input is not available from the beginning, we use an online algorithm as in [23, 44]. The total achieved data rate of each UE when assigning it on different CCs' RBs, i.e., r_i, requires the knowledge of $\phi_{i,z_{k,j}}$ on each RB $z_{k,j}$ the UE is assigned on. We use an online scheduling algorithm to decrease the computation overhead while processing the rate information as in [23].

Let $\phi_{i,z_{k,j}}[n]$ be the proportion of the frames that UE i is scheduled on RB $z_{k,j}$ in the first n frames. Then, the proportion of the frames that UE i is scheduled on RB $z_{k,j}$ in the $[n+1]$th frame is defined as follows:

$$\phi_{i,z_{k,j}}[n+1] = \begin{cases} \frac{n-1}{n} \phi_{i,z_{k,j}}[n] + \frac{1}{n}, \\ \text{if UE } i \text{ is scheduled on RB } z_{k,j} \\ \text{in the } (n+1)\text{th frame} \\ \frac{n-1}{n} \phi_{i,z_{k,j}}[n], \text{otherwise.} \end{cases} \tag{7.5}$$

In the proposed scheduling policy, for certain CC's RB $z_{k,j}$, the eNodeB schedules the UE that maximizes $\frac{U_i'(c_{i,f_k}+\sum_{z_{k,j}\in\mathscr{Z}_k}\phi_{i,z_{k,j}}H_{i,z_{k,j}})H_{i,z_{k,j}}}{U_i(c_{i,f_k}+r_{i,f_k})}$ on RB $z_{k,j}$.

Lemma 7.3.1 *Using the scheduling policy in (7.5), we show that* $\liminf_{n\to\infty}\sum_{i=1}^{M_k}\log U_i(c_{i,f_k}+\sum_{z_{k,j}\in\mathscr{Z}_k}(\phi_{i,z_{k,j}}[n]H_{i,z_{k,j}}))$ *exists for optimization problem (7.4).*

Proof We define $L(\phi)=\sum_{i=1}^{M_k}\log U_i(c_{i,f_k}+\sum_{z_{k,j}\in\mathscr{Z}_k}(\phi_{i,z_{k,j}}H_{i,z_{k,j}}))$ where ϕ, $\phi[n]$ and H are the short terms for $\phi_{i,z_{k,j}}$, $\phi_{i,z_{k,j}}[n]$ and $H_{i,z_{k,j}}$, respectively. Let $r_{i,f_k}[n]=\sum_{z_{k,j}}(\phi_{i,z_{k,j}}[n]H_{i,z_{k,j}})$. Using Taylor's theorem, for any ϕ and $\Delta\phi$ we have

$L(\phi+\Delta\phi)=L(\phi)+L'(\phi)\Delta\phi+\pi(\phi,\Delta\phi)$

where $|\pi(\phi+\Delta\phi)|<b|\Delta\phi|^2$, for some constant b.

Let $\Delta\phi_{i,z_{k,j}}[n]=\phi_{i,z_{k,j}}[n+1]-\phi_{i,z_{k,j}}[n]$, then

$$\Delta\phi_{i,z_{k,j}}[n]=\begin{cases}\frac{1}{n}-\frac{\phi_{i,z_{k,j}}[n]}{n}, & \\ \text{if UE } i \text{ is scheduled on RB } z_{k,j} & \\ \text{in the } (n+1)\text{th frame} & \\ \frac{-\phi_{i,z_{k,j}}[n]}{n}, \text{ otherwise.}\end{cases}$$

$|\Delta\phi_{i,z_{k,j}}[n]|<\frac{1}{n}$, for all i and $z_{k,j}$. As a result;

$$\begin{aligned}L(\phi[n+1])&=L(\phi[n]+\Delta\phi[n]),\\&\geq L(\phi[n])+\Delta L(\phi[n])-\frac{b}{n^2},\\&=L(\phi[n])+\left(\sum_i\frac{U_i'(c_{i,f_k}+\sum_{z_{k,j}}\phi H)}{U_i(c_{i,f_k}+r_{i,f_k})}H\Delta\phi\right)-\frac{b}{n^2}\\&=L(\phi[n])+\frac{1}{n}\left(\max_i\frac{U_i'(c_{i,f_k}+\sum_{z_{k,j}}\phi H)}{U_i(c_{i,f_k}+r_{i,f_k})}H\right.\\&\quad\left.-\sum_i\frac{U_i'(c_{i,f_k}+\sum_{z_{k,j}}\phi H)}{U_i(c_{i,f_k}+r_{i,f_k})}H\phi[n]\right)-\frac{b}{n^2}\\&\geq L(\phi[n])-\frac{b}{n^2},\end{aligned}\tag{7.6}$$

where $\Delta\phi[n]$ is substituted by $(\frac{1}{n}-\frac{\phi_{i,z_{k,j}}[n]}{n})$ (i.e., user i has the largest $\frac{U_i'(c_{i,f_k}+\sum_{z_{k,j}}\phi H)H}{U_i(c_{i,f_k}+r_{i,f_k})}$ among all users) and the last inequality holds since $\sum_i\phi_{i,z_{k,j}[n]}=1$ for all i and $z_{k,j}$.

Let $\beta:=\limsup_{n\to\infty}L(\phi[n])$. For any $\varepsilon>0$, there exists large enough N so that $L(\phi[N])>\beta-\frac{\varepsilon}{2}$ and $\sum_{n=N}^{\infty}\frac{b}{n^2}<\frac{\varepsilon}{2}$. For any $\hat{n}>N$, $L(\phi[\hat{n}])\geq L(\phi[N])-\sum_{n=N}^{\hat{n}}\frac{b}{n^2}>\beta-\varepsilon$. Therefore, $L(\phi[n])$ converges to β, as $n\to\infty$.

Due to the constraint $\sum_{i=1}^{M_k} \phi_{i,z_{k,j}} = 1$ in (7.4), ϕ is a solution to optimization problem (7.4) if and only if

$$\begin{aligned} \frac{dL}{d\phi_{i,z_{k,j}}} &= \frac{U_i'(c_{i,f_k} + \sum_{z_{k,j}} \phi H)H}{U_i(c_{i,f_k} + r_{i,f_k})} \\ &= \max_m \frac{U_m'(c_{m,f_k} + \sum_{z_{k,j}} \phi_{m,z_{k,j}} H_{m,z_{k,j}})}{U_m(c_{m,f_k} + r_{m,f_k})} H_{m,z_{k,j}}, \end{aligned} \tag{7.7}$$

for all i and $z_{k,j}$ such that $\sum_{i=1}^{M_k} \phi_{i,z_{k,j}} = 1$ and $\phi_{i,z_{k,j}} \geq 0$.

Theorem 7.3.2 *Using the scheduling policy (7.7),* $\lim_{n\to\infty} L(\phi)[n] = \sum_{i=1}^{M_k} \log U_i (c_{i,f_k} + \sum_{z_{k,j}} (\phi_{i,z_{k,j}}[n] H_{i,z_{k,j}}))$ *(i.e.,* $\lim_{n\to\infty} L(\phi[n])$*) achieves the maximum of optimization problem (7.4).*

Proof Suppose $\lim_{n\to\infty} L(\phi[n])$ does not achieve the maximum of the optimization problem. There exists $\delta > 0$, $\lambda > 0$, and positive integer N such that for all $n > N$, there exists some $i^n \in M_k$ and $z_{k,j}^n \in \mathscr{Z}_k$ so that $\phi_{i^n, z_{k,j}^n}[n] > \delta$ and $\frac{U_{i^n}'(c_{i^n f_k} + \sum_{z_{k,j}} \phi_{i^n, z_{k,j}^n} H_{i^n, z_{k,j}^n}) H_{i^n, z_{k,j}^n}}{U_i(c_{i^n f_k} + r_{i^n f_k})} < \max_m \frac{U_m'(c_{m,f_k} + \sum_{z_{k,j}} \phi_{m,z_{k,j}^n} H_{m,z_{k,j}^n}) H_{m,z_{k,j}^n}}{U_m(c_{m,f_k} + r_{m,f_k})} - \lambda$. Now we have:

$$\begin{aligned} L(\phi[n+1]) - L(\phi[n]) &\geq L'(\phi[n]))\Delta\phi[n] - \frac{b}{n^2} \\ &= \sum_{i=1}^{M_k} \frac{U_i'(c_{i,f_k} + \sum_{z_{k,j}} \phi[n] H)H}{U_i(c_{i,f_k} + r_{i,f_k})} \Delta\phi[n] - \frac{b}{n^2} \\ &= \frac{\delta\lambda}{n} - \frac{b}{n^2} \geq \frac{\delta\lambda}{2n}, \end{aligned}$$

for large enough n. Since $\sum_{n=1}^{\infty} \frac{1}{n} = \infty$, which is a contradiction. As a result, $\lim_{n\to\infty} L(\phi[n])$ achieves the maximum of the optimization problem.

7.4 Centralized UPFRS-CA Optimization Algorithm

In this section, we present our centralized utility proportional fairness resource scheduling with carrier aggregation algorithm. Each UE is assigned RBs from all in rage component carriers. The proposed UPFRS-CA algorithm is used by the eNodeB to allocate the available component carriers' RBs to users subscribing for mobile services based on a proportional fairness policy. First, the eNodeB performs the user grouping method described in Sect. 7.2 and creates one user group set $\mathscr{M}_{f_k}$ for all users under the coverage of component carrier f_k. It then allocates RBs of each component carrier starting with carrier f_1 (i.e., the carrier with the smallest

Algorithm 7 eNodeB UPFRS-CA Algorithm

Initialize $\phi_{i,z_{k,j}} = 0$, $r_{i,f_k} = 0$; $r_i = 0$; $c_{i,f_1} = 0$
Receive application utility parameters k_i, a_i and b_i from all UEs under the coverage area of the eNodeB
Create user groups $\mathcal{M}_{f_1}, \mathcal{M}_{f_2}, \ldots, \mathcal{M}_{f_K}$
for $k = 1 \rightarrow K$ **do**
  for $j = 1 \rightarrow |\mathcal{Z}_k|$ **do**
    Estimate the channel gain $G_{i,z_{k,j}}$ $\forall i \in \mathcal{M}_{f_k}$
    Calculate $H_{i,z_{k,j}}$ $\forall i \in \mathcal{M}_{f_k}$
    if $m = \arg\max_i \frac{U_i'(c_{i,f_k}+\phi_{i,z_{k,j}}H_{i,z_{k,j}})H_{i,z_{k,j}}}{U_i(c_{i,f_k}+r_{i,f_k})}$ **then**
      Allocate RB $z_{k,j}$ to the mth UE in $\mathcal{M}_{f_k}$
      $\phi_{m,z_{k,j}}[n+1] = \frac{n-1}{n}\phi_{i,z_{k,j}}[n] + \frac{1}{n}$
      $\phi_{i,z_{k,j}}[n+1] = \frac{n-1}{n}\phi_{i,z_{k,j}}[n]$ {For $i \neq m$}
    end if
    Calculate $r_{i,f_k} = \sum_{z_{k,q}=z_{k,1}}^{z_{k,j}}(\phi_{i,z_{k,q}}H_{i,z_{k,q}})$ $\forall i \in \mathcal{M}_{f_k}$
  end for
  if $k < K$ **then**
    $c_{i,f_{k+1}} = \sum_{l=1}^{k} r_{i,f_l}$
  end if
end for
Allocate $r_i = \sum_{l=1}^{K} r_{i,f_l}$

coverage radius) in ascending order $f_1 \rightarrow f_K$. For each component carrier f_k, the eNodeB allocates each of the f_k RBs, i.e., $z_{k,j}$, to the UE that has the maximum $\frac{U_i'(c_{i,f_k}+\phi_{i,z_{k,j}}H_{i,z_{k,j}})H_{i,z_{k,j}}}{U_i(c_{i,f_k}+r_{i,f_k})}$. Users with sigmoidal-like utility functions representing their real-time applications are given priority when allocating RBs due to the nature of their applications. Once the eNodB is done assigning component carrier f_k RBs, it then calculates the next component carrier $c_{i,f_{k+1}}$ value, for each user, which represents the aggregated rate allocated to user i from component carriers $f_1, \ldots, f_k$. The same process is repeated until the eNodeB assigns carrier f_K RBs and allocates each user with its final aggregated rate $r_i = \sum_{l=1}^{K} r_{i,f_l}$. Algorithm 7 shows the eNodeB centralized UPFRS-CA algorithm.

7.5 Numerical Results

In this section we present simulation results for the proposed resource scheduling with CA approach. We consider an LTE-Advanced mobile system with $M = 8$ users and two CCs f_1 and f_2 available at the eNodeB with $f_1 > f_2$ as shown in Fig. 7.1. We apply the user grouping method presented in Sect. 7.2 and two user groups are obtained, $\mathcal{M}_{f_1} = \{1, 2, 3, 4\}$ and $\mathcal{M}_{f_2} = \{1, 2, \ldots, 8\}$ where user $i \in \mathcal{M}_{f_k}$ represents the ith user located under the coverage area of carrier f_k. Users $\{1, 2, 5, 6\}$ are running real-time applications that are represented by sigmoidal-like utility functions with parameters $a_i = 5$ and $b_i = 10$ for users $\{1, 5\}$ and

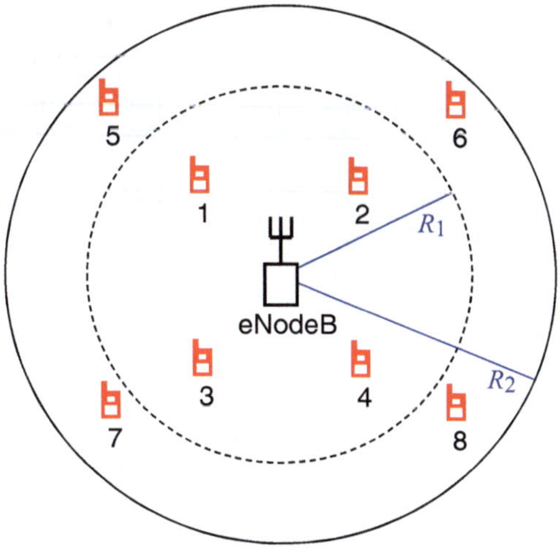

Fig. 7.1 LTE-Advanced mobile system with two component carriers (i.e., f_1 and f_2) available at the eNodeB with $f_1 > f_2$ and coverage radius $R_1 < R_2$. $\mathcal{M}_{f_1} = \{1, 2, 3, 4\}$, $\mathcal{M}_{f_2} = \{1, 2, \ldots, 8\}$ represent the sets of user groups located under the coverage area of carrier f_1 and f_2, respectively

$a_i = 1$ and $b_i = 30$ for users $\{2, 6\}$. Users $\{3, 4, 7, 8\}$ are running delay-tolerant applications that are represented by logarithmic utility functions with parameters $k_i = 15$ for users $\{3, 7\}$ and $k_i = 0.5$ for users $\{4, 8\}$. The simulation was run using MATLAB (7.7).

We compare the performance of the resource scheduling with CA approach in the case of using the proposed utility proportional fairness (UPF) resource scheduling policy and in the case of using the traditional proportional fairness (traditional-PF) scheduling policy presented in [23]. We assume equal channel gain in our simulation. In Fig. 7.3, we show simulation results and compare the performance of different scheduling policies for users in $\mathcal{M}_{f_1}$ that are assigned RBs by carrier f_1 and users in $\mathcal{M}_{f_2}$ that are assigned RBs by carrier f_1 and f_2. Figure 7.2 shows users' applications quality of experience (QoE) with their assigned RBs from carrier f_1 and carrier f_2.

Figure 7.3 shows the objective function of carrier f_1 RA optimization problem that is given by the multiplication of all users' applications quality of experience (QoE) for users in $\mathcal{M}_{f_1}$ and the objective function of carrier f_2 RA optimization problem when using the aforementioned scheduling policies. Figure 7.3 shows that the system performance, represented by the objective function value of the RA optimization problem that is given by the multiplication of all users applications' utilities, that represent users' satisfaction with the allocated rates in the case of the proposed UPF scheduling policy, is much greater than the objective function value when using the traditional-PF scheduling policy. It also shows that the system performance when using the traditional-PF with equal priority weights is worse than the system performance when using the traditional-PF with non-equal priority weights.

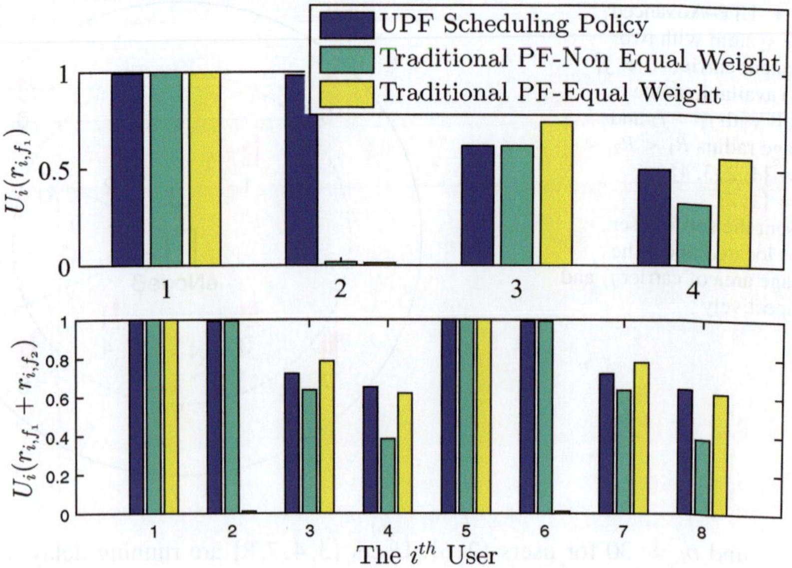

Fig. 7.2 Users' applications quality of experience comparison for different scheduling policies represented by the applications' utility functions of the assigned RBs from carrier f_1 and f_2

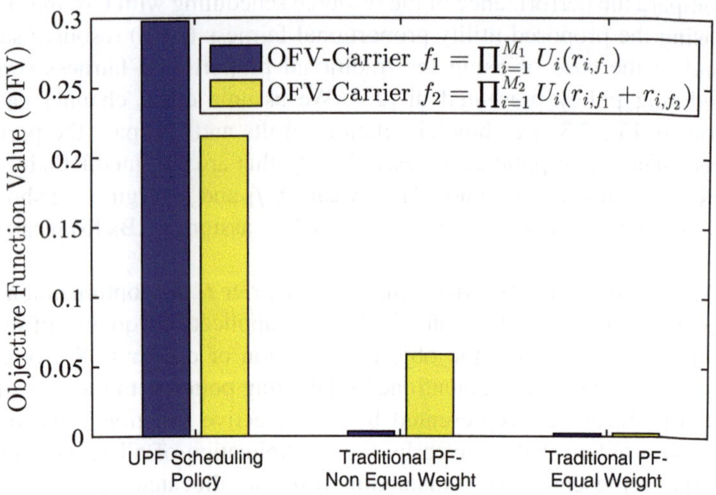

Fig. 7.3 Performance comparison for different scheduling policies represented by the objective function of carrier f_1 and f_2 RA optimization problems

7.6 Summary and Conclusion

In this chapter, we introduced a RB scheduling with CA approach in LTE-Advanced. Users are partitioned in user groups and each user is assigned on RBs of its corresponding in range carriers. We used utility PF with CA policy and presented users' applications using utility functions. We proved that our scheduling policy exists and therefore the optimal solution is tractable. Simulation results showed that the proposed resource scheduling with CA policy achieves better QoE than the traditional proportional fairness policy.

7.7 MATLAB Code

```
%1. Main Code
function UE
close all
clear all
clc
%%Two component carriers C1 and C2 with resources available
    at the eNodeB%%
%%%%%%%%%%%%%%Initialize%%%%%%%%%%%%%%%%%%%
Phi_C1 = [0 0 0 0 0 0 0 0]; %phi for the users that belong to
    user group 1, i.e under the coverage area of C1
Phi_C2 = [0 0 0 0 0 0 0 0]; %phi for the users that belong to
    user group 2, i.e under the coverage area of C2
Phi_C1_new = [0 0 0 0 0 0 0 0];
Phi_C2_new = [0 0 0 0 0 0 0 0];

a = [5 1];
b = [10 30];
r_i=[1 1 1 1 1 1 1 1];

r_i_C1 = [1 1 1 1 1 1 1 1];
r_i_C2 = [1 1 1 1 1 1 1 1];

c_i_C1 = [0 0 0 0 0 0 0 0];
c_i_C2 = [0 0 0 0 0 0 0 0];

H_i = 1;
Sum1 = [0 0 0 0 0 0 0 0];
Sum2 = [0 0 0 0 0 0 0 0];

%%%%%%%%%% RB Scheduling for Carrier 1 %%%%%%%%%%
for RB = 1:60
    %%%%%%%%%% Group 1 of users %%%%%%%%%%
    j = RB;
    for i = 1: 4
```

```
            Sum1(1:4) = j;
            Phi_C1_new(1:4) = ((j-1)./j).*Phi_C1(1:4)+(1./j);
            [soln11(i), soln1(i)] = SchedFunc1(i, RB, c_i_C1(1:4), r_i_C1(1:4), Sum1(1:4), Phi_C1_new(1:4));
            Val1(i)= soln1(i);
        end
        [M,I] = max(Val1);

        for i = 1:4
            if i == I
                Phi_C1(i) = Phi_C1_new(i);
                r_i_C1(i) = Sum1(i)*Phi_C1(i);
            else
                Phi_C1(i) = ((j-1)./j).*Phi_C1(i);
                r_i_C1(i) = Sum1(i)*Phi_C1(i);
            end
        end
end

c_i_C2(1:4) = r_i_C1(1:4) %rates allocated to users of group 1
utility_C1 = soln11
System_utility_C1 = prod(utility_C1)

%%%%%%%%%% RB Scheduling for Carrier 2 %%%%%%%%%
for RB2 = 1:100
        %%%%%%%%%% Group 2 of users %%%%%%%%%
        j = RB2;
        for i = 1: 8
            Sum2(1:8) = j;
            Phi_C2_new(1:8) = ((j-1)./j).*Phi_C2(1:8)+(1./j);
            [soln22(i),soln2(i)] = SchedFunc2(i, RB2, c_i_C2(1:8), r_i_C2(1:8), Sum2(1:8), Phi_C2_new(1:8));
            Val2(i)=soln2(i);
        end
        [M2,I2] = max(Val2);

        for i = 1:8
            if i == I2
                    Phi_C2(i) = Phi_C2_new(i);
                    r_i_C2(i) = Sum2(i)*Phi_C2(i);
            else
            Phi_C2(i) = ((j-1)./j).*Phi_C2(i);
            r_i_C2(i) = Sum2(i)*Phi_C2(i);
            end
        end
end
r_i_C2
%% Aggregated rate allocated to users from carrier 1 and carrier 2
TotalRare_C1_C2 =c_i_C2 + r_i_C2
utility_C1_C2 = soln22
System_utility_C1_C2 = prod(utility_C1_C2)
```

```
%2. Called by Main
function [R,T] = SchedFunc1(ii,RB,c_i,r_i_C1,sum,Phi_C1_new)

f = [0 0 0 0 0 0 0 0];
k = [15 0.5];
a = [5 1];
b = [10 30];
c = (1+exp(a.*b))./(exp(a.*b));
d = 1./(1+exp(a.*b));
x = c_i + (sum .*Phi_C1_new);
gamma = c_i + (sum .*Phi_C1_new);

for i = 1: 2
    m(i) = exp(-a(i).*(x(i)-b(i)));
    m1(i) = exp(-a(i).*(gamma(i)-b(i)));
    dy_sig(i) = a(i).*m(i)./((1+m(i)).*(1-d(i).*(1+m(i))));
    y_sig(i) = c(i).*((1./(1+m1(i)))-d(i));
    dy_log(i + 2) = k(i)./((1+k(i).*x(i+2)).*log(1+k(i).*100)
        );
    y_log(i + 2) = (log(1+k(i).*gamma(i+2)))./ log(1+k(i)
        .*100);
    dy(i) = dy_sig(i);
    dy(i + 2) = dy_log(i + 2);
    y(i) = y_sig(i);
    y(i + 2) = y_log(i + 2);
end

T = dy(ii)./y(ii);
R = y(ii);

%3. Called by Main
function [R2,T2] = SchedFunc2(ii,RB2,c_i,r_i_C2,sum,
    Phi_C2_new)

f = [0 0 0 0 0 0 0 0];
k = [15 0.5];
a = [5 1];
b = [10 30];
c = (1+exp(a.*b))./(exp(a.*b));
d = 1./(1+exp(a.*b));
x = c_i + (sum .*Phi_C2_new);
gamma = c_i + (sum .*Phi_C2_new);

for i = 1: 2
    m(i) = exp(-a(i).*(x(i)-b(i)));
    m1(i) = exp(-a(i).*(gamma(i)-b(i)));
    dy_sig(i) = a(i).*m(i)./((1+m(i)).*(1-d(i).*(1+m(i))));
    y_sig(i) = c(i).*((1./(1+m1(i)))-d(i));
    dy_log(i + 2) = k(i)./((1+k(i).*x(i+2)).*log(1+k(i).*100)
        );
    y_log(i + 2) = (log(1+k(i).*gamma(i+2)))./ log(1+k(i)
        .*100);
    dy(i) = dy_sig(i);
    dy(i + 2) = dy_log(i + 2);
```

```
        y(i) = y_sig(i);
        y(i + 2) = y_log(i + 2);
    end
    for i = 5: 6
        m(i) = exp(-a(i-4).*(x(i)-b(i-4)));
        m1(i) = exp(-a(i-4).*(gamma(i)-b(i-4)));
        dy_sig(i) = a(i-4).*m(i)./((1+m(i)).*(1-d(i-4).*(1+m(i)))
            );
        y_sig(i) = c(i-4).*((1./(1+m1(i)))-d(i-4));
        dy_log(i + 2) = k(i-4)./((1+k(i-4).*x(i+2)).*log(1+k(i-4)
            .*100));
        y_log(i + 2) = (log(1+k(i-4).*gamma(i+2)))./ log(1+k(i-4)
            .*100);
        dy(i) = dy_sig(i);
        dy(i + 2) = dy_log(i + 2);
        y(i) = y_sig(i);
        y(i + 2) = y_log(i + 2);

    end

    T2 = dy(ii)./y(ii);
    R2 = y(ii);
```

References

1. H. Shajaiah, A. Abdelhadi, C. Clancy, Utility proportional fairness resource allocation with carrier aggregation in 4g-lte, in *IEEE Military Communications Conference (MILCOM)*, 2013
2. G. Yuan, X. Zhang, W. Wang, Y. Yang, Carrier aggregation for LTE-advanced mobile communication systems. IEEE Commun. Mag. **48** 88–93 (2010)
3. Z. Shen, A. Papasakellariou, J. Montojo, D. Gerstenberger, F. Xu, Overview of 3GPP LTE-advanced carrier aggregation for 4G wireless communications. IEEE Commun. Mag. **50**, 122–130 (2012)
4. Frequency Spectrum Wall Chart (Commerce Dept., National Telecommunications and Information Administration, Office of Spectrum Management, 2016)
5. M. Awad, V. Mahinthan, M. Mehrjoo, X. Shen, J.W. Mark, A dual-decomposition-based resource allocation for OFDMA networks with imperfect CSI. IEEE Trans. Veh. Technol. **59**, 2394–2403 (2010)
6. M. Mehrjoo, S. Moazeni, X.S. Shen, Resource allocation in OFDMA networks based on interior point methods. Wirel. Commun. Mob. Comput. **10**(11), 1493–1508 (2010)
7. P. Tejera, W. Utschick, J. Nossek, G. Bauch, Rate balancing in multiuser MIMO OFDM systems. IEEE Trans. Commun. **57**, 1370–1380 (2009)
8. A. Abdelhadi, C. Clancy, A utility proportional fairness approach for resource allocation in 4G-LTE, in *IEEE International Conference on Computing, Networking, and Communications (ICNC), CNC Workshop* (2014)
9. A. Abdelhadi, C. Clancy, A robust optimal rate allocation algorithm and pricing policy for hybrid traffic in 4G-LTE, in *IEEE nternational Symposium on Personal, Indoor, and Mobile Radio Communications (PIMRC)* (2013)
10. A. Abdelhadi, C. Clancy, J. Mitola, A resource allocation algorithm for multi-application users in 4G-LTE, in *MobiCom Workshop* (2013)

11. H. Shajaiah, A. Abdelhadi, C. Clancy, A price selective centralized algorithm for resource allocation with carrier aggregation in ltc cellular networks, in *2015 IEEE Wireless Communications and Networking Conference (WCNC)*, March 2015, pp. 813–818
12. H. Shajaiah, A. Khawar, A. Abdel-Hadi, T. Clancy, Resource allocation with carrier aggregation in LTE Advanced cellular system sharing spectrum with S-band radar, in *2014 IEEE International Symposium on Dynamic Spectrum Access Networks (DYSPAN)*, April 2014, pp. 34–37
13. A. Abdelhadi, C. Clancy, An optimal resource allocation with joint carrier aggregation in 4G-LTE, in *2015 International Conference on Computing, Networking and Communications (ICNC)*, February 2015, pp. 138–142
14. Executive Office of the President, President's Council of Advisors on Science and Technology (PCAST), *Realizing the Full Potential of Government-Held Spectrum to Spur Economic Growth* (PCAST, Washington, DC, 2012)
15. M. Richards, J. Scheer, W. Holm, *Principles of Modern Radar* (Scitech, Raleigh, 2010)
16. Federal Communications Commission (FCC), In the matter of revision of parts 2 and 15 of the commission's rules to permit unlicensed national information infrastructure (U-NII) devices in the 5 GHz band. MO&O, ET Docket No. 03–122, June 2006
17. Federal Communications Commission, Proposal to create a Citizen's Broadband Service in the 3550–3650 MHz band (2012)
18. Federal Communications Commission (FCC), Connecting America: The national broadband plan. Online (2010)
19. NTIA, An assessment of the near-term viability of accommodating wireless broadband systems in the 1675–1710 mhz, 1755–1780 mhz, 3500–3650 mhz, 4200–4220 mhz and 4380–4400 mhz bands (2010)
20. National Telecommunications and Information Administration (NTIA), Analysis and resolution of RF interference to radars operating in the band 2700–2900 MHz from broadband communication transmitters. Online, October 2012
21. M. Cotton, R. Dalke, Spectrum occupancy measurements of the 3550–3650 megahertz maritime radar band near San Diego, California (2014)
22. T. Erpek, A. Abdelhadi, C. Clancy, An optimal application-aware resource block scheduling in LTE, in *IEEE International Conference on Computing, Networking and Communications (ICNC) Workshop CCS)* (2015)
23. I.-H. Hou, C.S. Chen, Self-organized resource allocation in LTE systems with weighted proportional fairness, in *2012 IEEE International Conference on Communications (ICC)*, June 2012, pp. 5348–5353
24. H. Shajaiah, A. Abdelhadi, T.C. Clancy, Towards an application-aware resource scheduling with carrier aggregation in cellular systems. IEEE Commun. Lett. **20**, 129–132 (2016)
25. I. Toufik, S. Sesia, M. Baker, *LTE - The UMTS Long Term Evolution: From Theory to Practice* (Wiley, New York, 2011)
26. G. Tychogiorgos, A. Gkelias, K.K. Leung, Utility-proportional fairness in wireless networks, in *PIMRC* (IEEE, Piscataway, 2012), pp. 839–844
27. Y. Wang, A. Abdelhadi, A QoS-based power allocation for cellular users with different modulations, in *2016 International Conference on Computing, Networking and Communications (ICNC)*, February 2016, pp. 1–5
28. A. Abdelhadi, H. Shajaiah, Optimal resource allocation for smart phones with multiple applications with MATLAB instructions. Technical Report (2016)
29. A. Abdelhadi, H. Shajaiah, Application-aware resource allocation with carrier aggregation using MATLAB. Technical Report (2016)
30. Z. Kbah, A. Abdelhadi, Resource allocation in cellular systems for applications with random parameters, in *2016 International Conference on Computing, Networking and Communications (ICNC)*, February 2016, pp. 1–5
31. A. Abdelhadi, H. Shajaiah, Optimal resource allocation for cellular networks with MATLAB instructions. CoRR, abs/1612.07862 (2016)

32. F. Wilson, I. Wakeman, W. Smith, Quality of service parameters for commercial application of video telephony, *in Proceedings of the Human Factors in Telecommunications Symposium* (Darmstadt, 1993)
33. S. Shenker, Fundamental design issues for the future internet. IEEE J. Sel. Areas Commun. **13**, 1176–1188 (1995)
34. Y. Wang, A. Abdelhadi, T.C. Clancy, Optimal power allocation for lte users with different modulations, in *2016 Annual IEEE Systems Conference (SysCon)*, April 2016, pp. 1–5
35. M. Ghorbanzadeh, A. Abdelhadi, C. Clancy, Distributed resource allocation, in *Cellular Communications Systems in Congested Environments* (Springer, Cham, 2017), pp. 61–91
36. J.-W. Lee, R.R. Mazumdar, N.B. Shroff, Downlink power allocation for multi-class wireless systems. IEEE/ACM Trans. Networking **13**, 854–867 (2005)
37. A. Kumar, A. Abdelhadi, T.C. Clancy, A delay efficient multiclass packet scheduler for heterogeneous M2M uplink, in *IEEE MILCOM* (2016)
38. A. Kumar, A. Abdelhadi, T.C. Clancy, An online delay efficient packet scheduler for M2M traffic in industrial automation, in *IEEE Systems Conference* (2016)
39. A. Kumar, A. Abdelhadi, T.C. Clancy, A delay optimal MAC and packet scheduler for heterogeneous M2M uplink. CoRR, abs/1606.06692 (2016)
40. A. Kumar, A. Abdelhadi, T.C. Clancy, A delay-optimal packet scheduler for M2M uplink, in *IEEE MILCOM* (2016)
41. R. Madan, S. Boyd, S. Lall, Fast algorithms for resource allocation in wireless cellular networks. IEEE/ACM Trans. Networking **18**, 973–984 (2010)
42. J.B. Taylor, *Principles of Microeconomics*. Microeconomics Series (Houghton Mifflin, Boston, 1998)
43. S. Boyd L. Vandenberghe, *Introduction to Convex Optimization with Engineering Applications*. Course Reader (Orlando, 2001)
44. T. Erpek, A. Abdelhadi, T.C. Clancy, Application-aware resource block and power allocation for LTE, in *2016 Annual IEEE Systems Conference (SysCon)*, April 2016, pp. 1–5

Chapter 8
Conclusion and Future Trajectory

In this chapter, we summarize the contributions provided in this book and discuss some possible research directions in the future to improve and expand the proposed methods presented in this book.

In this book, we introduced new optimal resource allocation methods for future wireless systems that take into consideration aggregating multiple wireless providers' resources. We showed the efficiency of the proposed methods compared to other existing methods in improving mobile users' quality of experience. The resource allocation with CA problem is formulated into a convex optimization framework based on utility proportional fairness where the RA optimization problem gives priority to real-time applications due to the nature of the sigmoidal-like utility functions. Both distributed and centralized multi-stage resource allocation with CA algorithms are presented to allocate multi-carrier resources optimally among users running real-time and delay-tolerant applications. In addition, a resource allocation with user discrimination framework is proposed to allocate single or multiple carriers resources among users running multiple applications. Furthermore, we developed resource allocation with carrier aggregation methods and algorithms for spectrum sharing systems that involve commercial use of under-utilized spectrum. Multi-stage resource allocation with CA algorithms are proposed for sharing the Federal under-utilized 3.5 GHz spectrum with commercial users. In addition, we presented a spectrum sharing scenario between radar and communication systems and developed optimal resource allocation with CA algorithm to allocate their resources. Additionally, a resource block scheduling with carrier aggregation for multi-carrier cellular systems is proposed to allocate multiple component carriers' RBs to active UEs based on utility proportional fairness policy and carrier aggregation. We proved that the proposed scheduling policy provides tractable optimal solution. Simulation results showed that the proposed resource scheduling method outperforms traditional scheduling mechanisms as it provides better applications QoE.

H. Shajaiah et al., *Resource Allocation with Carrier Aggregation in Cellular Networks*, DOI 10.1007/978-3-319-60540-1_8

An outline of future research direction is as follows:

- Develop resource allocation with carrier aggregation framework for heterogeneous radio access technologies such as LTE-A and WiFi. Such approach is a challenging one because the two technologies have differences in spectrum access and physical layer. However, the coexistence of LTE and WiFi provides benefit for both, it increases the rate capacity for LTE end users and in the same time introduces reliability for WiFi users. This requires resource management algorithms that are based on CA where UEs are connected to LTE eNodeB as well as WiFi access point and are allocated resources from both, based on carrier aggregation.
- Develop efficient component carriers assignment algorithms for networks incorporating LTE-A with Carrier Aggregation. Component carriers involved in the resource assignment process should not be assigned based on the signal received power level, estimated by each UE, as it is insufficient in case of carrier aggregation. It is more important to consider the load on each component carrier which can be estimated from counting users on each CC or by measuring the interference on each CC. For example, in heterogeneous networks, the coexistence of macro cells and pico cells makes the resource management task challenging. This is due to the differences in output power between macro BS and pico BS; i.e., 43–46 dBm in the case of macro cell and 30 dBm in the case of pico cell, which causes under-utilization in small cells' BS resources due to their lower power levels that make most of users select macro BS.
- Consider providing methods to mitigate the interference caused by carrier aggregation. For example, in LTE-A network, when considering femto cells that are installed by customers, it is important to develop a scheme that optimizes the carrier aggregation selection while avoiding interference among eNodeBs and femto cells. This can be achieved by selecting an optimal group of component carriers for carrier aggregation in an LTE-A network.
- There are other problems that can benefit from the simulation tools provided in this book. For example, machine communications (M2M) in [1–4], multi-cast network [5], ad-hoc network [6–9], and other wireless networks [10–13].

References

1. A. Kumar, A. Abdelhadi, T.C. Clancy, A delay efficient multiclass packet scheduler for heterogeneous M2M uplink, in *IEEE MILCOM* (2016)
2. A. Kumar, A. Abdelhadi, T.C. Clancy, An online delay efficient packet scheduler for M2M traffic in industrial automation, in *IEEE Systems Conference* (2016)
3. A. Kumar, A. Abdelhadi, T.C. Clancy, A delay optimal MAC and packet scheduler for heterogeneous M2M uplink. CoRR, abs/1606.06692 (2016)
4. A. Kumar, A. Abdelhadi, T.C. Clancy, *Design and Implementation of Practical Schedulers for M2M Uplink Networks Using MATLAB* (Springer International Publishing, Cham, 2017)
5. A. Abdel-Hadi, S. Vishwanath, On multicast interference alignment in multihop systems, in *IEEE Information Theory Workshop 2010 (ITW 2010)* (2010)

6. H. Zhou, X. Wang, Z. Liu, X. Zhao, Y. Ji, S. Yamada, "Qos-aware resource allocation for multicast service over vehicular networks," *2016 8th International Conference on Wireless Communications Signal Processing (WCSP)*, October 2016, pp. 1–5
7. Z. Fan, Y. Li, G. Shen, C.C.K. Chan, Dynamic resource allocation for all-optical multicast based on sub-tree scheme in elastic optical networks, in *2016 Optical Fiber Communications Conference and Exhibition (OFC)*, March 2016, pp. 1–3
8. J. Jose, A. Abdel-Hadi, P. Gupta, S. Vishwanath, On the impact of mobility on multicast capacity of wireless networks, in *2010 Proceedings IEEE INFOCOM*, March 2010, pp. 1–5
9. S. Gao, M. Tao, Energy-efficient resource allocation for multiple description coding based multicast services in OFDMA networks, in *2016 IEEE/CIC International Conference on Communications in China (ICCC)*, July 2016, pp. 1–6
10. S. Chieochan, E. Hossain, Downlink media streaming with wireless fountain coding in wireline-cum-WiFi networks. Wirel. Commun. Mob. Comput. **12**, 1567–1579 (2012)
11. A. Abdelhadi, F. Rechia, A. Narayanan, T. Teixeira, R. Lent, D. Benhaddou, H. Lee, T.C. Clancy, Position estimation of robotic mobile nodes in wireless testbed using GENI. CoRR, abs/1511.08936 (2015)
12. S. Chieochan, E. Hossain, Wireless fountain coding with IEEE 802.11e block ACK for media streaming in wireline-cum-WiFi networks: a performance study. IEEE Trans. Mob. Comput. **10**, 1416–1433 (2011)
13. S. Chieochan, E. Hossain, Network coding for unicast in a WiFi hotspot: promises, challenges, and testbed implementation. Comput. Netw. **56**, 2963–2980 (2012)

Appendix: C++ Functions

```
// Below is the code to set the size of the console window

void SetWindow(int Width, int Height)
    {
    _COORD coord;
    coord.X = Width;
    coord.Y = Height;

    _SMALL_RECT Rect;
    Rect.Top = 0;
    Rect.Left = 0;
    Rect.Bottom = Height - 1;
    Rect.Right = Width - 1;

    HANDLE Handle = GetStdHandle(STD_OUTPUT_HANDLE);
    // Get Handle
    SetConsoleScreenBufferSize(Handle, coord);
    // Set Buffer Size
    SetConsoleWindowInfo(Handle, TRUE, &Rect);
    // Set Window Size
    }

//--------------------------------------------------------//
//Logarithmic utility function
float U_Log(float r,float k_i){
    float exp = 2.718281828459;
float U2 = (log(1+(k_i*r)))/(log(1+(k_i*(100)))) ;
```


H. Shajaiah et al., *Resource Allocation with Carrier Aggregation in Cellular Networks*, DOI 10.1007/978-3-319-60540-1

```
return (U2);
}
//Sigmoidal-like utility function
float U_Sig(float r,float a_i,float b_i){
    float exp = 2.718281828459;
float U1 = ((1+(pow(exp,a_i*b_i)))/(pow(exp,a_i*b_i))) *
    ((1/(1+(pow(exp,-a_i*(r-b_i)))))-(1/(1+(pow(exp,a_i*b_i)))
    ));
return (U1);
}

//--------------------------------------------------------------//
//f(r) Functions:
float F1(float r,float a_i,float b_i,float p,float opt1,
    float opt2){
    float exp = 2.718281828459;
    float res1 =((a_i*(1/pow(exp,a_i*b_i))*(pow(exp,-a_i*(r+
        opt1+opt2-b_i)))) / (1-(1/(1+(pow(exp,a_i*b_i))))*(1+(
        pow(exp,-a_i*(r+opt1+opt2-b_i))))) + ((a_i*pow(exp,-
        a_i*(r+opt1+opt2-b_i)))/(1+(pow(exp,-a_i*(r+opt1+opt2-
        b_i))))))-p;
    return (res1);
}


float F2(float r,float k_i,float p,float opt1, float opt2){
    float res2 =(k_i/((1+(k_i*(r+opt1+opt2)))*(log(1+(k_i
    *(100))))))-p;
    return (res2);
}

float F1_disc(float r,float a_i,float b_i,float p,float opt,
float a1)
{
    float exp = 2.718281828459;
    float res1 =a1*((a_i*(1/pow(exp,a_i*b_i))*(pow(exp,-a_i*(
        r-b_i)))) / (1-(1/(1+(pow(exp,a_i*b_i))))*(1+(pow(exp
        ,-a_i*(r-b_i))))) + ((a_i*pow(exp,-a_i*(r-b_i)))/(1+(
        pow(exp,-a_i*(r-b_i))))))-p;
    return (res1);
}

float F2_disc(float r,float k_i,float p,float opt,float a2)
{
    float res2 =a2*(k_i/((1+(k_i*(r)))*(log(1+(k_i*(100))))))
```

```
    -p;
    return (res2);
}

float F3_disc(float r,float k_i,float a_i,float b_i,float p,
    float opt1,float opt2,float a1,float a2,float B){
    float exp = 2.718281828459;
    float res3 =B*(a1*((a_i*(1/pow(exp,a_i*b_i))*(pow(exp,-
        a_i*((r+opt1)-b_i)))) / (1-(1/(1+(pow(exp,a_i*b_i))))
        *(1+(pow(exp,-a_i*((r+opt1)-b_i))))) + ((a_i*pow(exp,-
        a_i*((r+opt1)-b_i)))/(1+(pow(exp,-a_i*((r+opt1)-b_i)))
        ))) + a2*((k_i/((1+(k_i*(r+opt1)))*(log(1+(k_i*(100)))
        )))))-p;
    return res3;
}

//------------------------------------------------------------//
//Root Functions use the Bisection method to solve equation
//f(r) = 0
float get_roots1(float num1,float num2,float p,float opt1,
float opt2,float R){
        float mid;
        float val;
        float low=.001;
        float high=R;
        for(int i = 0; i < 1000; ++i){
        mid = (low + high)/2;
        float val = F1(mid,num1,num2,p,opt1,opt2);
        if(val < 0.001 && val > -0.001){
        break;
        }
        else if (val < 0 )
        high = mid;
        else
        low = mid;
        }
        return (mid);
}

float get_roots2(float num,float p,float opt1,float opt2,
float R){
        float mid;
        float val;
        float low=.001;
        float high=R;
```

```
        for(int i = 0; i < 1000; ++i){
        mid = ((low + high)/2);
        float val = F2(mid,num,p,opt1,opt2);
        if(val < 0.001 && val > -0.001){
        break;
        }
        else if (val < 0 )
        high = mid;
        else
        low = mid;
        }
        return (mid);
}
//-----------------------------------------------------//
float get_roots1_disc(float num1,float num2,float p,float
opt,float R,float a1){
        float mid;
        float val;
        float low=.001;
        float high=R;
        for(int i = 0; i < 1000; ++i){
                mid = (low + high)/2;
                float val = F1_disc(mid,num1,num2,p,opt,a1);
                if(val < 0.001 && val > -0.001){
                        break;
                }
                else if (val < 0 )
                        high = mid;
                else
                        low = mid;
        }
                return (mid);
}

float get_roots2_disc(float num,float p,float opt,float R,
float a2){
        float mid;
        float val;
        float low=.001;
        float high=R;
        for(int i = 0; i < 1000; ++i){
                mid = ((low + high)/2);
                float val = F2_disc(mid,num,p,opt,a2);
                if(val < 0.001 && val > -0.001){
                        break;
```

```
                }
            else if (val < 0 )
                    high = mid;
             else
                    low = mid;
        }
        return (mid);
}

float get_roots3_disc(float num,float a_i,float b_i,float p,
float opt1,float opt2,float a1,float a2,float R,float B){
        float mid;
        float val;
        float low=.001;
        float high=R;
        for(int i = 0; i < 1000; ++i){
                mid = ((low + high)/2);
                float val = F3_disc(mid,num,a_i,b_i,p,opt1,
                opt2,a1,a2,B);
                if(val < 0.001 && val > -0.001){
                        break;
                }
            else if (val < 0 )
                    high = mid;
             else
                    low = mid;
        }
        return (mid+opt1+opt2);
}

//-------------------------------------------------------//
//This function returns the price per unit resource//
float shadowPrice(float arr1[],float Rate,int N_users){
    float sum= 0.0,price;
    for(int i = 0; i<N_users; i++){
        sum += arr1[i];
    }
    price = sum/Rate;
    return price;
}
```

Index

Symbols
3.5 GHz band, 133-149, 153, 189
3550–3650 MHz, 2, 153, 154

A
Access point, 190
Application-aware, 8, 133
Application status differentiation, 75
Applications usage percentages, 86, 87
Application target rate, 64–77, 79, 105, 110
Application utility function, 39, 74, 77, 90, 134, 175
Application weight, 75, 77, 84, 105
Asynchronous transfer mode (ATM), 3

C
Centralized resource allocation, 10, 11, 38–48
Channel-selection, 11, 154, 156–158, 169
Coefficient of reflection, 156
Commercial users, 2, 10, 11, 22, 63–75, 105, 109, 110, 133, 134, 142, 153, 189
Complex unitary matrix, 158
Computation overhead, 177
Convex, 10, 68, 78, 89
Convex optimization, 10, 26–28, 38, 40, 42, 66–68, 78–80, 96, 98, 136, 137, 160, 161, 177, 189
Council of Advisers on Science and Technology, 2, 4, 173
Coverage radius, 90–92, 98, 99, 101, 134, 176, 180, 181

D
Delay-tolerant applications, 7, 11, 20, 32, 37–39, 46, 48, 64, 69, 71, 73, 75, 77, 83, 85, 88, 89, 99, 102, 104, 105, 135, 140, 153, 154, 163, 164, 166, 168–170, 189
Distributed resource allocation algorithm, 30–31

E
Emergency services, 2, 88
Euclidean norm, 157
Evolve node B (eNodeB), 1, 8, 10, 22, 25–28, 30, 31, 34, 36–48, 64–71, 75–83, 85–96, 98–105, 133–139, 141–145, 148, 153, 154, 161–165, 167, 173–174, 176–181, 183, 190

F
Federal Communications Commission (FCC), 4, 153, 173
Fluctuation decay function, 46, 84, 164
Frank Kelly algorithm, 21
FTP, 20

G
Global optimal solution, 10, 26–28, 40, 42, 65, 67, 68, 76, 78–80, 89, 96, 98, 105, 136, 137, 160, 161, 177

H. Shajaiah et al., *Resource Allocation with Carrier Aggregation in Cellular Networks*, DOI 10.1007/978-3-319-60540-1

H
Heterogeneous network (HetNet), 8, 133, 134, 190
High traffic, 86, 87

I
Implementation, 2, 6–8, 175
IMT-Advanced, 1, 4
Inelastic traffic, 3, 4, 8, 10, 35, 48, 64, 67, 73, 74, 79, 133, 173
Inflection point, 19, 20, 33–35, 71, 73, 74, 99, 166
Inter-band non contiguous, 1, 6, 7
Interference channel, 8, 9, 153–155, 157–159, 169
Intra-band contiguous, 6
Intra-band non contiguous, 6, 7

J
Joint users, 38, 39, 45, 46, 48

L
Lagrange multipliers, 29, 40, 42, 98, 136, 138
Lagrangian, 28, 29, 40, 42, 98, 136, 137
Leased resources, 134, 135
Logarithmic utility, 19, 20, 22, 23, 25, 26, 32–34, 37–39, 45, 50, 51, 53, 54, 64, 65, 69, 71, 77, 88, 96, 97, 99, 109, 111, 134, 135, 140, 153, 163, 165, 167, 170, 175, 181, 193

M
Macro cell, 2, 6, 133–139, 142, 143, 145, 147, 190
MIMO radar, 8, 9, 11, 153–157, 159–166, 168, 169
Minimum degradation, 169
Minimum QoS, 7, 10, 21, 26, 38, 48, 64, 75, 91, 105, 106, 170, 176
Minimum required utility, 133–136, 140, 142, 146
Multi-application resource allocation, 75–88
Multi-carrier systems, 8, 38–39, 173
Multi-flow CA, 8
Multi input multi output (MIMO), 8, 9, 153, 154
Multimedia telephony, 1, 64
Multi-stage resource allocation, 10, 25–61, 88, 89, 91, 173, 189

N
National Telecommunications and Information Administration (NTIA), 4, 173
Null-space computation, 11, 154, 156, 158
Null-space projection (NSP), 11, 154, 157–159

O
Offered price, 25, 38–43, 45–48
Open Systems Interconnection (OSI), 3
Optimal aggregated rate, 161

P
Path loss, 4, 156, 176
Price sensitivity, 86–88, 169
Priority weights, 181
Propagation loss, 156
Proportional fairness, 3, 7, 20, 21, 38, 40, 48, 65, 75, 79, 85, 89, 98, 136, 154, 160, 165, 173, 179, 181, 183
Public safety users, 2, 22, 63–75, 88, 104, 105, 109, 110

Q
QoS requirement, 2, 64, 176
Quality of service (QoS), 3, 4, 7, 9, 10, 20, 27, 174, 177

R
Radar, 2, 4, 8, 9, 11, 153, 154, 156, 157, 159, 162–165, 167–170, 189
Radar waveform, 9, 157–159
Real-time application, 7, 10, 20, 26–28, 33–35, 38–40, 42, 45, 48, 64–66, 69, 71, 75, 77, 83–85, 89, 91, 99, 102, 104–106, 136, 140–142, 153, 160, 161, 163–166, 168, 170, 175–177, 180, 189
Regular users, 77, 85, 88–93, 99, 102, 103, 106
Resource block allocation, 8
Resource block scheduling, 8, 11, 173–186, 189
Robust algorithm, 11
Router, 3

S
S-band Radar, 12, 153–170
Scheduling policy, 11, 12, 174, 178, 179, 181–183, 189
Shadow price, 29–31, 40, 42–45, 48, 52, 54, 68–75, 80–82, 86–88, 98, 104, 105, 107, 108, 136, 138, 161–163, 169

Sigmoidal utility, 10, 19, 20, 22, 25, 26, 32–35, 38, 39, 45, 46, 50, 51, 53, 64, 65, 68, 69, 71, 72, 77, 83, 88, 91, 97, 99, 109–111, 134, 135, 140, 141, 153, 160, 163–165, 167, 170, 173, 175, 176, 180, 189, 193
Signal to noise ratio (SNR), 175
Single carrier, 7, 10, 26–28, 36, 88, 173
Single-flow CA, 8
Singular value decomposition (SVD), 158
Small cell, 2, 8, 133–143, 145, 146, 148, 190
Solution existence, 105
Steering matrix, 156
Strictly concave, 7, 19, 26, 39, 68, 77, 96–98, 159, 177
Subscriber differentiation, 64, 75
System model, 32, 38–39, 45, 76, 101, 135, 154–155, 157, 174–175

T
Taylor's theorem, 178
Third-generation partnership project (3GPP), 2–4, 6
Traffic-dependant, 104
Transmission overhead, 38, 48
Transmit antennas, 154
Transmit steering vector, 156

U
Under-utilized spectrum, 2, 134, 135, 189
Uplink, 6
User discrimination, 10, 11, 22, 63–130, 189
User equipment (UE), 1, 8, 10, 22, 26, 30–32, 37, 41–43, 45, 64–71, 73, 75–83, 85, 86, 88, 89, 91, 92, 98–100, 105, 113, 133, 138, 140, 142, 153–155, 159, 161–163, 170, 173–180, 190
User grouping method, 89–91, 98, 174–176, 179, 180
Utility function, 7, 10, 11, 19–23, 25, 26, 32–35, 37–39, 45, 46, 48, 50, 51, 53, 54, 64–66, 68, 69, 71–77, 83, 84, 88–91, 93, 96–99, 101, 105, 107–111, 133–135, 139–141, 153, 159, 160, 163–167, 170, 173–177, 180–183, 189, 193
Utility proportional fairness, 7, 8, 10, 11, 20–22, 27, 30, 66, 67, 91–95, 105, 134, 136, 137, 141, 160, 170, 173–186, 189

V
Video streaming, 20
VIP users, 75, 82, 83, 85, 88–94, 99, 102–104, 106
VoIP, 20

W
Weighted fair queuing, 3
WiFi, 190
WiMAX, 3
Wireless local area network (WLAN), 8

MIX
Papier aus verantwortungsvollen Quellen
Paper from responsible sources
FSC® C105338

FSC
www.fsc.org

If you have any concerns about our products,
you can contact us on
ProductSafety@springernature.com

In case Publisher is established outside the EU,
the EU authorized representative is:
Springer Nature Customer Service Center GmbH
Europaplatz 3, 69115 Heidelberg, Germany

Printed by Libri Plureos GmbH
in Hamburg, Germany